Shuja ur Rehman
Shoaib Hasan
Muhammad Mujahid Iqbal

Emerging Trends in Chemistry, Physics, Biochemistry, Plant Sciences

Shuja ur Rehman
Shoaib Hasan
Muhammad Mujahid Iqbal

Emerging Trends in Chemistry, Physics, Biochemistry, Plant Sciences

Noor Publishing

Imprint

Cover image: www.ingimage.com

Publisher:
Noor Publishing
is a trademark of
Dodo Books Indian Ocean Ltd. and OmniScriptum S.R.L publishing group

120 High Road, East Finchley, London, N2 9ED, United Kingdom
Str. Armeneasca 28/1, office 1, Chisinau MD-2012, Republic of Moldova, Europe
Managing Directors: Ieva Konstantinova, Victoria Ursu
info@omniscriptum.com

Printed at: see last page
ISBN: 978-620-8-87085-0

Preface

The contents of the book explaining the scientific ideas. It is comprehensively explored the principles of the natural sciences, chemistry, biochemistry, plant and physics. The natural and biological sciences are emerging fields revealing the nature for novel information. From healthcare and technology to the environmental protection, they influence the many parts of human existence by providing the basic information about the physical universe, living things, and their relationships.

MAIN CONTENTS

Chapter 1 Emerging Trends in Medical Biochemistry

Introduction to Authors

Shuja ur Rehman working as a Assistant Professor and HOD Biochemistry and MLT Department, Saint Mary's College of Medical and Advance Studies Gujranwala. He has contributed to this book chapter as the first author.

Shoaib Hasan is working as a Lecturer at the Institute of Molecular Biology and Biotechnology (IMBB), The University of Lahore. His research experience spans across Biotechnology, Enzymology, and Biomedical Science. He has contributed to this book chapter as the first author.

1.1 Artificial Intelligence in Biochemistry

Deep learning is a distinct category of machine learning that employs multi-layered artificial neural networks. The layered approach allows deep learning models to acquire hierarchical feature representations from unstructured input data. Deep learning excels in analyzing extensive, complex biological data due to its ability to autonomously identify significant patterns.

The architecture and functionality of the artificial neural networks are based on the biological networks. They consist of strata of artificial neurons interconnected by weighted links. Each neuron receives input from other neurons, processes that input, and then communicates its output to further neurons, facilitating the flow of information across the network.

1.2 Artificial Intelligence techniques in Biochemistry

Deep learning algorithms can predict and assess metabolic and signalling networks by integrating proteomics and metabolomics data with existing pathway databases. This facilitates the clarification of the fundamental causes of illnesses and the identification of prospective therapeutic targets. AI-driven route analysis may pinpoint essential regulatory nodes, reveal intricate network connections, and elucidate the impact of exploration on biological processes. To effectively use artificial intelligence in biology, it is essential to build technology capable of autonomously gathering biological data from many sources, including photos, videos, and molecular profiles. Data scientists must collaborate with biologists to guarantee the reliability and accuracy of the data. Nonetheless, the quality of the gathered data is as significant.

Three primary issues must be tackled in forthcoming endeavours: algorithmic openness, data quality and curation, and ethical concerns. Integrating these advanced computational tools is important for the future of biochemistry research, and the prospects offered by this technology are too significant to overlook. As the capabilities of AI in biochemistry research are further investigated, it is crucial to acknowledge that the primary objective is to enhance human health and well-being. The capacity of AI to comprehend the intricate systems that underpin life may facilitate the creation of more effective medications and interventions to tackle the most urgent health issues facing world today. This will enhance the comprehension of biological processes.

The objective of every machine learning model is to discern patterns in the training data that may be used on new, unexamined data. A broadly generalized model may nevertheless forecast outcomes on novel data, even if that data differs from the training data in some aspects. Robust generalization is essential for developing models that are applicable in the real-world scenarios. Conversely, overfitting transpires when a model gets too intricate, aligning with the noise in the training data instead of the fundamental patterns. As a result, a model may demonstrate exceptional performance on training data while performing inadequately on new, unexamined data.

To prevent overfitting and enhance generalization, machine learning practitioners use methods such as cross-validation, regularization, and early stopping.

- Utilizations
- Diagnostic Assessment
- Image examination
- Oncology
- cardiovascular disease
- Predictive analytics
- Customized Therapy
- Pharmaceutical development
- Robotic-assisted surgery
- Patient surveillance

- Optimized Operations
- Automation of administrative tasks
- Optimization of research
- clinical investigations

1.3 Methods in Biochemistry

Molecular biology techniques are employed to examine the structure, function, and modification of proteins, RNA, DNA, and other biological macromolecules. The comprehension of the gene expression, cellular interactions, and organismal development and behaviour relies on these techniques.

The polymerase chain reaction (PCR) is a prevalent laboratory method for the generating numerous copies (millions or billions) of a specific DNA segment.

This DNA region may represent any subject of interest to the researcher. It may be a genetic marker utilized by forensic experts to correlate suspects DNA with that obtained from crime scenes, or it could be a gene that a researcher aims to examine. The objective of PCR is often to generate a sufficient quantity of the target DNA region for analysis or other applications. For example, PCR-amplified DNA may be submitted for sequencing, analyzed through gel electrophoresis, or cloned into a plasmid for additional investigation.

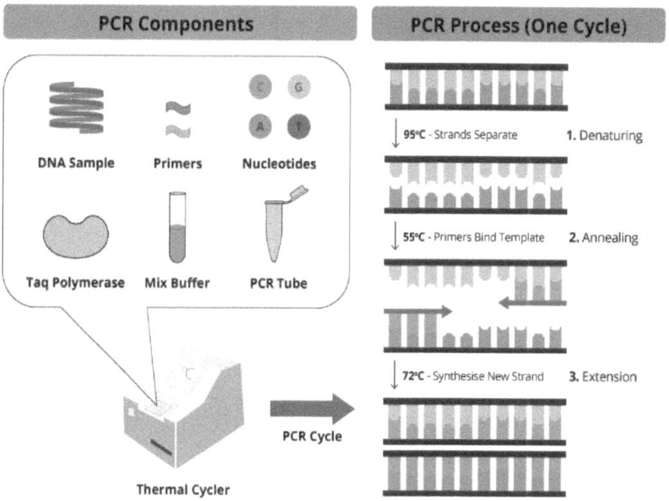

Fig 1.1 Shows the different steps of PCR reaction

Fundamental elements

- Deoxyribonucleic acid template
- Collection of primers
- Taq polymerase enzyme
- deoxyribonucleotides (dNTPs)
- reaction buffer
- distilled water

Polymerase Chain Reaction (PCR), an essential element of DNA replication in an organism, necessitates the DNA polymerase enzyme to the synthesize new DNA strands utilizing existing strands as the templates. The DNA polymerase commonly utilized in PCR is referred to as Taq polymerase, named after the thermophilic bacterium from which it was derived (Thermus aquaticus). *T. aquaticus* inhabits hot springs and hydrothermal vents.

Its DNA polymerase exhibits optimal activity at 70 °C, a temperature at which neither *E. coli* nor human DNA polymerase would operate, owing to its remarkable thermal stability. The thermal stability of *Taq polymerase* renders it optimal for PCR applications. High temperatures are typically employed in PCR to denature, or separate, the strands of the template DNA.

Each PCR reaction employs two primers, with each primer strategically placed to encompass the target region designated for amplification. In other words, they receive sequences that facilitate binding to the complementary strands of the template DNA, precisely at the locus designated for replication.

Primary Steps

- Elevate the procedure to a high temperature to denature or dissociate the DNA strands. This offers a single-stranded template for the subsequent step.
- Cool the reaction to facilitate the binding of primers to their complementary sequences on the single-stranded template DNA.
- Increasing the reaction temperature will enhance Taq polymerase's ability to elongate the primers and generate more DNA strands.

Diverse Applications

- pharmacology
- forensic science
- Research in molecular biology
- medical diagnostics
- ecological domains
- microorganisms that cause disease
- hereditary disorders
- oncogenic mutations

1.4 Electrophoresis

The outcomes of a PCR reaction are often seen by gel electrophoresis. The gel electrophoresis technique utilizes an electric current to propel DNA fragments across a gel matrix, facilitating their separation based on size. A standard, sometimes called a DNA ladder, is usually supplied to determine the size of the fragments in the PCR sample. A DNA band has several copies of the target DNA region instead of only one or a few. DNA must exist in substantial quantities to be visible to the naked eye due to its minuscule size.

An electric field facilitates migration of negatively charged nucleic acid molecules across a gel matrix composed of agarose, polyacrylamide, or an alternative medium for the purpose of separation.

Shorter molecules traverse greater distances and velocities than longer molecules due to their enhanced ability to navigate through the gel's pores. This occurrence is referred to as sieving. The charge in agarose facilitates the separation of proteins, since the gel's pores are sufficiently big to allow their passage. In gel electrophoresis studies, the gel functions as an anticonvective or sieving medium.

Gels may mitigate heat convection induced by the electric field and maintain the integrity of the separation for subsequent post-electrophoresis staining.

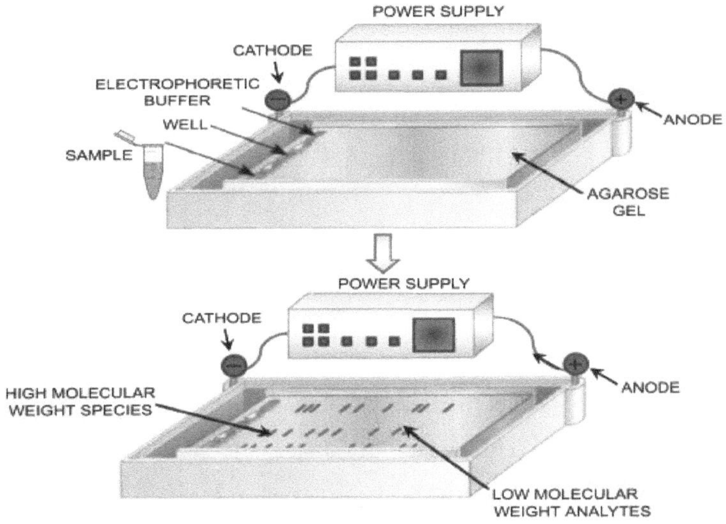

Fig 1.2 Shows the electrophoresis process illustrating the steps

The electrophoresis process enables the arrangement of molecules according on their size, shape, or charge. DNA and other molecules may traverse an agarose or polyacrylamide gel when subjected to an electric field. A positive charge at one end of the electric field attracts molecules through the gel, while a negative charge at the other end repels them. A cavity in the gel medium is occupied by the molecules undergoing separation.

The gel is housed inside an activated the electrophoresis chamber. Smaller molecules traverse the gel more rapidly in the presence of an electric field, whereas larger molecules migrate more slowly.

Elements

- a gel matrix such as agarose or polyacrylamide
- buffers
- an electrical power source
- chemical substances
- loading dyes
- blemishes

Agarose gels

Natural polysaccharide polymers extracted from seaweed are used to produce agarose gels. Agarose gels are simpler to prepare and handle than other matrices because they undergo physical changes throughout the gelation process rather than chemical alterations. Furthermore, samples may be obtained easily. The gel may be stored in a plastic bag in the refrigerator once the experiment concludes.

Polyacrylamide gels

Polyacrylamide gels provide the uniform pore dimensions, allowing for the separation of the proteins ranging from 5 to 2,000 kDa by polyacrylamide gel electrophoresis (PAGE). Modifying the quantities of acrylamide and bis-acrylamide powder, as well as the duration of polymerization, allows for the regulation of pore size in the gel.

The gel molecules may stained for visibility after the completion of electrophoresis. Silver stain or Coomassie brilliant blue dye serves to detect proteins, whereas ethidium bromide, which fluoresces upon intercalation into DNA, is used for seeing DNA. The constituents of the mixture may also be seen separating on the gel by other approaches.

An autoradiogram of the DNA sequencing gel may be obtained if the molecules to be sorted are radioactive. Gels may be captured via photography, often with a gel documentation technique. On the prominent figure-creation platform SciUGo, gels are routinely annotated for scientific documentation and presentations.

Forensic genetics refers to the use of scientific methods and genetic instruments in criminal and civil litigation. Locard's Exchange posits that every transaction leaves a trace, making every evidence crucial for forensic inquiry.

Biological evidence, including cellular debris or cell-free DNA, may be present at crime scenes. As genetic technology has advanced, these procedures have been used for genetic study of both non-human animals and humans. Human DNA typing is the preferred method owing to the accessibility of the databases and established protocols, however same techniques may be used to any genome.

Applications

- molecular biology
- biochemistry

- forensic science
- Pharmaceuticals
- DNA profiling
- Polymerase Chain Reaction (PCR) analysis
- purification of proteins
- studies of gene expression

1.5 Nanotechnology

Nanotechnology refers to the manipulation of matter with the dimensions ranging from one to one hundred nanometers (nm). The surface area and quantum mechanical processes are essential in defining the properties of materials at the nanoscale. This definition of nanotechnology encompasses all technological and scientific methods related to these distinctive traits. The phrases "nanotechnologies" or "nanoscale technologies" in the plural are often used to denote research and applications that possess the characteristic of scale. The precise technical aim of manipulating atoms and molecules to create macroscale objects was an early definition of nanotechnology, currently referred to as molecular nanotechnology.

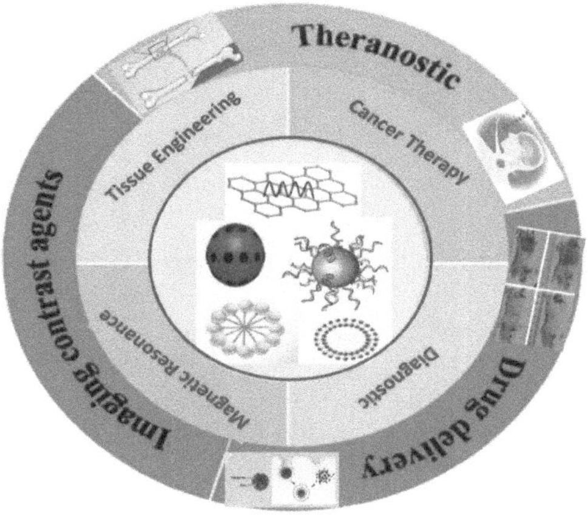

Fig 1.3 Shows the role of nanotechnology in the various fields

Principles

The top-down strategy necessitates phased construction of nanodevices, analogous to the industrial items. A fundamental technique for synthesis and characterization is scanning probe microscopy. Atoms may be manipulated and surfaces analyzed via scanning tunnelling and atomic force microscopes. These microscopes may be used to shape surfaces and direct self-assembling entities by establishing specific places. Atoms or molecules on a surface may be manipulated using scanning probe microscopy techniques, including the feature-oriented scanning method.

Clinical applications

Therapeutic drug delivery is prominent use of nanomedicine, whereby pharmaceuticals are administered using nanoparticles that function as vehicles to transport the drugs to appropriate site. The dimensions, morphology, surface charge, and surface modifications (such as proteins, coatings, and polymers) of nanoparticle carriers, which may consist of synthetic or organic components, may be altered to enhance their functionality.

The capacity to functionalize nanoparticles is especially advantageous when aiming at body regions with unique physiochemical properties that hinder the direct delivery of the desired medicine. Certain nanoparticles are capable of transporting therapeutic medications to the brain.

Chapter 2 Role of Computer Science and Artificial Intelligence for Technological Advancement

Introduction to authors of book chapter

1-Noman Javed from University of West Scotland, London with particularly research interest in Data Science for enhancement in big data handling and its applications

2-Awais Muzafar from Beijing Institute of Graphic Communication, China with particularly research interest in Artificial intelligence for enhancement in technology and applications in routine.

3-Jahangeer Ali from Department of Software Engineering, University of Engineering and Technology Taxila, Pakistan. To explore the respective field of Artificial Intelligence, Data Science, new technologies, and emerging trends in Artificial Intelligence and Computer Technology for the sake of learning and growth. As a research scholar, under the supervision of world-renowned researchers as well as in their respective labs, research projects and institutes, where I fully utilize my mental and physical potential to pursue my career in research and academia in field.

2.1 Current advances in Computer Science

Field of the computer science encompasses both theoretical and practical aspects of computing, information, and automation. It includes fields such as algorithm theory, theory of computation, and information theory, as well as fields that deal with the design and implementation of hardware and software.

Computer science relies heavily on the data structures and algorithms. Theorising computation involves creating abstract models of computation and identifying broad groups of issues that may be addressed using these models.

The study of safe communication methods and the prevention of security flaws is at the heart of computer security and cryptography. Creating visuals is the focus of computer graphics and computational geometry. Database theory deals with the administration of data repositories, whereas programming language theory explores various approaches to characterise computer operations. While software engineering is concerned with the theory and practice of creating software, human-computer interaction studies ways in which people and computers communicate with one another. The design and concepts of complicated systems are studied in areas including embedded systems, operating systems, and networks. The design of hardware and software that make up computers is known as computer architecture. The objective of artificial intelligence and machine learning is to replicate the ability to learn, adapt to new environments, solve the problems, and make decisions, much like humans and other animals.

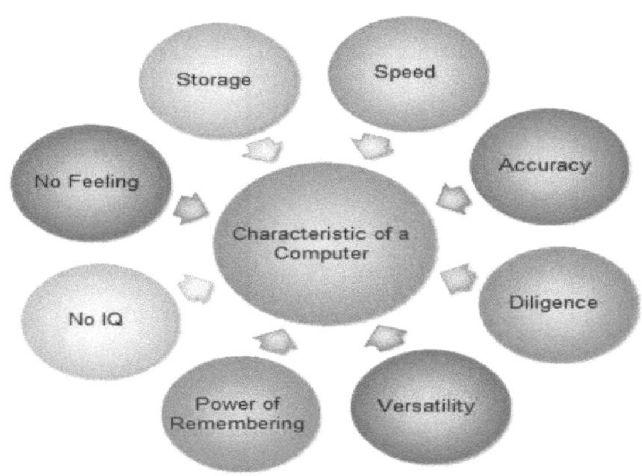

Fig 2.1 Shows the characteristics of computer and its illustrations

2.2 Current advances

Computer vision and natural language processing are two branches of artificial intelligence that seek to comprehend and analyse visual and auditory input. A wide variety of activities may be accomplished via the usage of programming languages. Popular models for computer programming consist of programming technique known as functional programming eliminates state and mutable data in favour of constructing computer programs in a way that views computing as the evaluation of mathematical functions. Programming in this paradigm is done using expressions or declarations rather than statements, it is a declarative programming paradigm.

A programming paradigm known as imperative programming employs statements to modify the state of a program. Statements in an imperative program instruct the computer to carry out certain tasks, similar to how the imperative mood in spoken languages conveys orders. The main goal of imperative programming is to define the behaviour of programs.

Object-oriented programming is a paradigm for creating computer programs that relies on the idea of "objects" to store information (in the form of fields, or attributes) and execute operations (in the form of methods, or code). One thing about objects is that their associated procedures may access and, in many cases, change the data fields of those objects. That is why interacting objects are the building blocks of object-oriented computer programs.

2.3 Principles components

The paradigm of the service-oriented programming involves the creating and implementing integrated business applications and mission-critical software systems using "services" as the unit of computer activity.

The process of creating and executing algorithms, which are detailed descriptions of operations, by writing code in one or more programming languages is known as computer programming or coding. Computers may then follow these programs to carry out various tasks.

Machine code is performed directly by the central processing unit, although programmers usually employ high-level languages that are more readily understood by humans. If there a be a good programmer, need to know a lot of things. Like the domain of the application, specific algorithms, formal logic, programming languages, and generic code libraries.

Along with the programming, there are auxiliary responsibilities including requirements analysis, testing, debugging (finding and correcting issues), build system implementation, and derived artefact management (such as programs' machine code). The words "coding," "implementation," and "programming" are reserved for the actual act of creating and altering code; nonetheless, the phrase "software development" is sometimes used to refer to this broader process. When software development uses formal methodologies or follows an engineering design process, it is often referred to as software engineering.

Fig 2.2 Shows the steps and characteristics of computer programming

How frequently a software produces accurate results is called its reliability. This is dependent on algorithms that are conceptually sound and avoid programming faults like off-by-one or division-by-zero blunders or resource management issues like buffer overflows and race situations.

Robustness is the degree to which a software can foresee issues caused by mistakes (rather than faults). Among them include user mistake, unanticipated power outages, memory, OS services, and network connection issues, as well as improper, unsuitable, or corrupt data.

What makes a software usable is its ergonomics, or how easily a human can use it for its intended or, in rare instances, unexpected, purposes. Regardless of other factors, they may determine its success or failure. The goal is to make the user interface of a software more clear, intuitive, coherent, and comprehensive using a variety of textual, visual, and sometimes hardware components.

A program's portability refers to the variety of computer hardware and operating system platforms that can build, understand, and execute the program's source code. This is conditional on the availability of platform-specific compilers (and occasionally libraries) for the language of the source code, as well as on the expected behaviour of the hardware and operating system, as well as on the programming facilities offered by the various platforms.

A program is considered maintainable if it can be easily changed by current or future developers to add features, customise it, correct security flaws and problems, or adapt it to other contexts. In this respect, good practices throughout early development are crucial. Even though this feature isn't immediately obvious to the user, it may have a major impact on a program's success or failure in the long run.

Program efficiency is a measure of how little system resources it uses. Less is better when it comes to CPU time, memory space, slow media like discs, network bandwidth, and even human engagement. This also involves meticulous resource management, such as removing unnecessary temporary files and memory leaks. A programming language tends to cast a shadow on these discussions.

Even slower languages, like Python, may run programs immediately from a human viewpoint, but the language does effect speed. Programs that slow down the system should prioritise speed, resource utilisation, and performance; however, other factors, such as cost (more hardware could be cheaper), and the effective use of programmer time should also be considered.

2.4 AI Function in Computer Science

An explosion of innovative opportunities has resulted from the fast development of artificial intelligence (AI), which is already pervasive in many facets of human existence. Generative pre-trained transformers (GPTs) have garnered a lot of attention, especially OpenAI's ChatGPT (GPT-3.5 version). The goal of these models is to make it easier to generate meaningful writing that sounds and looks like human-written material. The unique capabilities offered by all AI-based models are designed to automate certain jobs, improve productivity, and solve nebulous issues.

There are many parts of our life where artificial intelligence is already present, but one area where it is most noticeable is in the classroom. Artificial intelligence has the ability to revolutionise the way education is conducted. To make learning and teaching more entertaining, effective, and personalised might be a game-changer for both students and instructors.

Advanced learning systems that can tailor instruction to each student's unique needs and improve test scores have emerged with the growth of artificial intelligence. By enabling students to get immediate feedback and participate in highly customised learning environments, these technologies have the ability to revolutionise conventional approaches to education. Machine learning, for instance, is capable of creating presentations and homework. It may also help with class debates, queries, and individual comments. Thus, ChatGPT and other AI technologies might be great instructional aids.

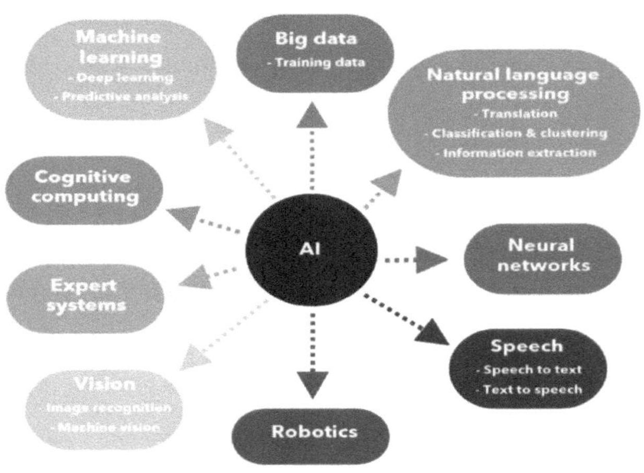

Fig 2.3 Shows role of AI and computer towards multidisciplinary approach

More and more, our everyday lives are being impacted by technology, AI, and the realm of computer science and technology in general. Computer science (CS) education is therefore critically important, and societal adaptation to new technologies is a requirement. Because it is a crucial set of skills, teaching computer science using technology is now mandatory for everyone.

Algorithms and data structures are two areas where students struggle while studying computer science. Also, not all students have the same level of development in their algorithmic mentality, which may make programming seem overwhelming. A large number of students have complained that some computer science classes are "dry" and "not very engaging".

One way that AI may improve computer science education is by enhancing existing teaching methods. Students may utilise these systems to help them with a variety of reasoning tasks, including those requiring both practical and abstract thinking. Also, students may learn to code more easily using ChatGPT since it can do so many different things.

2.5 Emerging aspects

With its own unique set of ideas and demanding standards for tackling complicated problems, database science (DB) is an important and separate branch of computer science. Query optimisation, relational database principles, and SQL syntax may be difficult for students to grasp, therefore new ways of teaching these concepts are needed. Improving the accessibility of learning abstract ideas might be achieved via the use of interactive learning techniques, real-time feedback, and explanations guided by artificial intelligence. Chatbots, which generate adaptive workouts and enable personalised learning paths, provide immediate assistance and are a potential AI solution.

There has been a dearth of study on the precise usage of chatbots and AI in computer science education, especially in database training, despite the abundance of papers investigating these topics generally.

So far, research on the pros, cons, and educational effects of using AI into database courses has been limited. To fill this need, this study presents a comprehensive analysis of artificial intelligence's (AI) influence on database learning as it pertains to computer science education. In order to guarantee the successful and pedagogically sound adoption of AI, it investigates its uses, benefits, drawbacks, and instructional design factors.

Simultaneously, it is impossible to ignore how AI has revolutionised programming. By automating crucial processes like coding, debugging, and testing, generative AI is changing the fundamental character of software creation, going beyond its function in education. It is because of this change that a new paradigm known as "AI-native software development" has emerged, in which programmers work less as coders and more as curators and collaborators with AI systems.

The way programming is taught must also be rethought in light of this revolution, which is redefining professional practices. To better equip students for a future when artificial intelligence (AI) capabilities are essential to growth, educational techniques will need to change.

By broadening the scope of the literature review to include new research on AI-assisted development, this study takes a closer look at how AI affects software engineering and the field as a whole, in addition to instructional design.

Despite its importance as a core ability in many fields, many students struggle with learning how to program. In basic computer science classes, problems may arise from many directions, including the students themselves, the teachers, and the very nature of the programming language itself.

To ensure that students not only create code that works but also follows proper conventions, it is important to use assessment rubrics that are designed for novice programmers. Standardising curricula and using programming languages with simple syntax can help with this.

A system or computer is said to have Artificial intellect (AI) if it can mimic human behaviours and intellect, including seeing, learning, and predicting. The education industry has been investigating the possibility of using AI for administrative purposes, to aid in teaching responsibilities, and to improve student learning.

One area of artificial intelligence that varies fundamentally from rule-based AI—which relies on explicit programming—is machine learning (ML). Machine learning (ML) enables computers to learn from data by creating suitable statistical models and algorithms. These algorithms then use this information to derive predictions or outcomes for new data points based on prior data. Artificial intelligence (AI) and machine learning (ML) have made great strides in two areas of educational research: educational data mining (EDM) and learning analytics (LA). In order to better understand both students and their learning environment, EDM centres on investigating and analysing educational data. Although LA makes use of some of the same methods as EDM, its primary goal is to enhance learning by means of data discovery, making it more learner-centred. Recent EDM has made use of Deep Learning (DL) for a number of tasks, including image recognition and natural language processing, since DL has become increasingly popular as a result of improvements in computing power.

2.6 Novel tools advancements

By raising the bar for data analysis, automation, and problem-solving, AI has utterly revolutionised the field of computer science. It has pushed computers to their limits, allowing for the creation of smart systems with intelligence, such as those that can learn, adapt, and make choices. Innovations in robotics, machine learning, and NLP have resulted from this.

Advancements in bioinformatics, autonomous systems, and big data analytics have been driven by artificial intelligence in computer science. Hence, AI has rethought the tech industry's skill sets and tool requirements while simultaneously improving computer systems' efficiency and capacities.

The fact is that we need programmers capable of creatively using AI, contrary to concerns that it will replace programmers. Despite AI's present usage in lowering software development costs, there is a greater need than ever before for programmers versed in AI.

Programming in the future will most likely include constant cooperation between humans and AI. Artificial intelligence (AI) teams outperform human-only teams in terms of productivity and effectiveness, according to recent research. So, although AI will keep displacing certain occupations, it will eventually create more employment as the partnership between humans and AI develops.

It know enough about AI to make any firm predictions about its future. It has already replaced numerous low-skill coding tasks, such creating boilerplate code, and it is still having a major influence on computer science progress. This is helpful since it reduces development time, which frees up developers to work on more intricate and imaginative parts of the project.

Nevertheless, there are still certain issues with AI, even if it is becoming better at coding and playing a part in the field of computer science. It generates code that isn't always efficient, introduces errors, and produces outcomes that still need professional debugging. So, learning to code is still useful in the modern digital world, but you'll need to supplement it with other talents if you want to keep up with all the new developments.

As far as the future of technology jobs is concerned, experts predict that the ability to solve complex problems at a high level will be as crucial as, if not more so than, the ability to code. Artificial intelligence has serious flaws that can only be addressed by humans, since it cannot evaluate subjective aspects like human motivation and relationships.

In the future, however, some worry that AI may lead to "social ruptures" amongst those who disagree with its sentience, particularly since "the dawn of consciousness in AI systems is likely by 2035," according to scholarly predictions. Companies are keen to use AI to save costs, but they should be mindful of the ethical and practical challenges that may arise.

Chapter 3 Current advances in Physics and Chemistry

Introduction to authors of book chapter

1-Muhammad Adnan is a Ph.D. scholar at the Shanghai Institute of Technical Physics, University of Chinese Academy of Sciences, Mainland China. His research focuses on investigating light–matter interactions in flexible two-dimensional (2D) materials using near-field optics. He is engaged in developing novel methods to engineer these interactions in van der Waals materials. His work includes the design and fabrication of photonic devices and the exploration of their applications in nanoscale light manipulation, materials engineering, quantum emission, imaging, sensing, energy transfer, and optical device development. He has contributed to this book chapter as the first author.

2-Salman Ali from School of Optoelctronics engineering, Xidian University of Science and Technology, China. Research interest: Single-frame infrared small target (SIRST) detection is a critical task in many applications, including military surveillance, search and rescue, and remote sensing. The objective is to accurately identify and isolate small targets in infrared images, often from cluttered and noisy backgrounds. However, the inherent characteristics of SIRST, such as low signal-to-noise ratio (SNR), small target size, and minimal contrast with the background, make this task extremely challenging.

3.1 Physics and its emerging features

Physics, the scientific study of the interplay between the most basic building blocks of visible cosmos and their physical properties. The scientific study of matter at all scales, from the macroscopic to the submicroscopic, is known as physics (from the Greek physikos). Not only does it investigate how things behave when subjected to certain forces, but it also delves into the history and composition of force fields like nuclear, electromagnetic, and gravitational. The final goal is to establish some all-encompassing principles that can explain and unify all these different occurrences.

From the most fundamental to the most advanced areas of study, physics holds the key. The fields of study that seek to identify and articulate the underlying principles of nature have historically been referred to as either physics or natural philosophy. This has changed very recently. Physics began to mean the branch of physical science that was not included by astronomy, chemistry, geology, or engineering as modern sciences evolved and became more specialised. Astrophysics, geophysics, biophysics, and even the psychophysics are subfields of the physics that focus on physical principles and observations; physics is fundamental to all the natural sciences. The trinity of matter, motion, and energy is the fundamental description of physics. The language of mathematics is used to convey its rules with economy and accuracy.

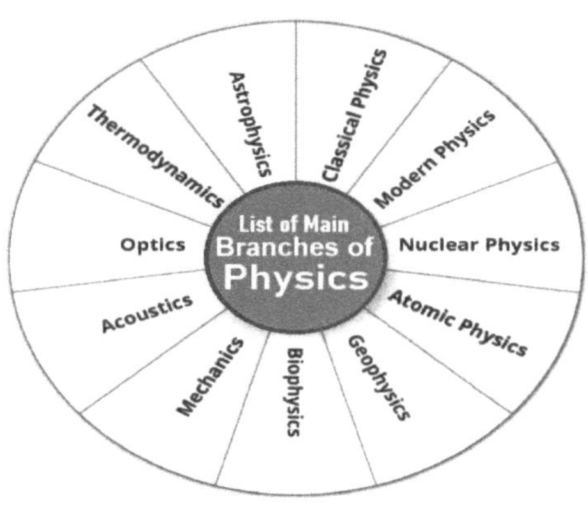

Fig 3.1 Shows the various domains of Physics and its associations

To further the field of physics, it is necessary to conduct experiments that are as controlled as possible and to formulate a coherent theoretical framework. These two approaches are complimentary but equally important. Measurements taken in physical experiments are then compared to the results anticipated theoretically.

A hypothesis is considered to embody a rule of physics if it consistently predicts the outcomes of relevant experiments. But a law may be changed, replaced, or confined to a smaller area at any time if a subsequent experiment dictates it.

Everybody in physics is working towards the same goal: a universal framework for explaining the behaviour of matter, energy, and motion at all scales, from the smallest (microscopic) subatomic particles to the biggest (extragalactic) scales. Much of this lofty objective has come to fruition. A remarkable collection of basic physical rules seems capable of explaining all observed occurrences, even if a fully integrated explanation of physical events has not been (and probably never will be) accomplished.

Classical physics, which was the dominant branch of the scientific community until about the 1900s, provides a reasonable explanation for the movements of huge, slow-moving objects in relation to the speed of light, as well as for the properties of heat, sound, electricity, magnetism, and light. When it comes to greater speeds, very big things, and the minuscule fundamental building blocks of matter, including electrons, protons, and neutrons, these principles are modified by the new breakthroughs in quantum mechanics and relativity.

3.2 Physical science methodology

Physics has changed and will change in the future without a set plan. Sophisticated measurements may uncover surprising behaviour in what is fundamentally an experimental field of study. Contrarily, aesthetic judgement, mathematical expansion of current ideas into new theoretical domains, critical re-examination of seemingly obvious but unproven assumptions, intuition, aesthetic evaluation, and comparison all play a part (as they do in all science). As an example, the quantum theory put out by the German scientist Max Planck relied on discrepancies between classical electromagnetic predictions and the observable properties of blackbody radiation. This kind of radiation is created by a heated body that absorbs all input radiation. A relativistic extension of the quantum theory of the electron was made by the English scientist P.A.M. Dirac, who predicted the existence of the positron. Instead of giving up on the conservation principles of beta decay, the German scientist Wolfgang Pauli proposed the illusive neutrino, which does not have mass or charge.

The electromagnetic theory of light was developed by Maxwell from his assumption that, given the well-established fact that changing electric fields can cause magnetic fields to change, he reasoned that the reverse would also be true. A revaluation of simultaneity became the basis of Albert Einstein's special theory of relativity, whereas the equivalence of inertial and gravitational mass is the foundation of his general theory of relativity.

While specific strategies may differ from one issue to another, a common goal among physicists is to reduce the complexity of previously intractable problems by simulating them as closely as possible in a succession of idealised models. Gas theory begins with the assumption that molecules are shapeless, infinitesimally tiny particles, similar to billiard balls. Iterative improvements are made to this perfect image.

Within the framework of quantum theory, the Danish physicist Niels Bohr developed the correspondence principle, a helpful guiding concept for expanding theoretical interpretations. In it, it is said that when a valid theory is expanded to cover more ground, the predictions of the expanded theory must align with those of the original theory within the shared domain where both theories apply.

When wave effects proportionate to the wavelength of light are minimal due to the smallness of that wavelength, for example, the more expansive theory of physical optics must have the same conclusion as the more limited theory of ray optics. When Planck's constant is tiny enough to be regarded negligible, the findings of classical and quantum mechanics must be identical. Also, at speeds that are negligible in comparison to the speed of light, such as baseballs in play, the laws of relativistic physics and Newtonian classical mechanics must be in agreement.

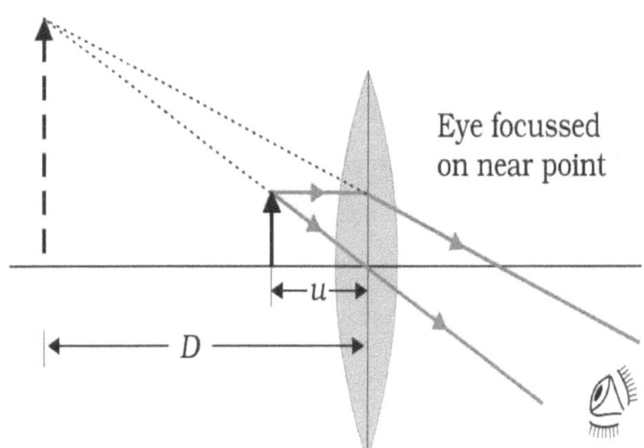

Fig 3.2 Shows the features characteristics of ray optics

The following examples show how experimental and theoretical physicists attempt to solve their challenges.

3.3 Emerging aspects

When new kinds of unstable particles, mostly high-energy protons from space, were detected in the atmosphere as a result of primary radiation, the contemporary experimental study of elementary particles could begin. Geiger detectors picked up on the new particles, while cloud chambers and photographic plates helped identify them. Particle physics, formerly known as high-energy nuclear physics, emerged as a prominent scientific discipline after WWII. High-energy particle accelerators of today may reach energies of billions of electron volts, cost hundreds of millions of dollars, and extend many kilometres in length.

Experimental teams, like the ones that found the W+, W−, and Z quanta of the weak force at CERN in Geneva, can consist of a hundred or more physicists from various countries, with a larger number of technical workers acting as support staff. These teams receive funding from the twenty member states of Europe. Particle physics labs make heavy use of the cutting-edge equipment, such as supercomputers and superconductive magnets, and use a wide range of visual and electronic methods to analyse and organise the massive volumes of data generated by their work.

What falls under the umbrella of "physics" are the fields that make up its many subfields. Physics is the scientific discipline that studies the nature and interplay of matter, energy, and their characteristics. Different subfields within physics have developed into full-fledged academic disciplines in recent decades. There are several subfields within physics, but the focus of classical physics is on the relationship between matter and energy. Classical mechanics, optics, audio, and electromagnetics are the four main areas of classical physics. It is becoming more impossible to categorise physics within the aforementioned domains due to the subject's exponential growth in recent decades. What follows is an overview of some of the most important subfields in physics.

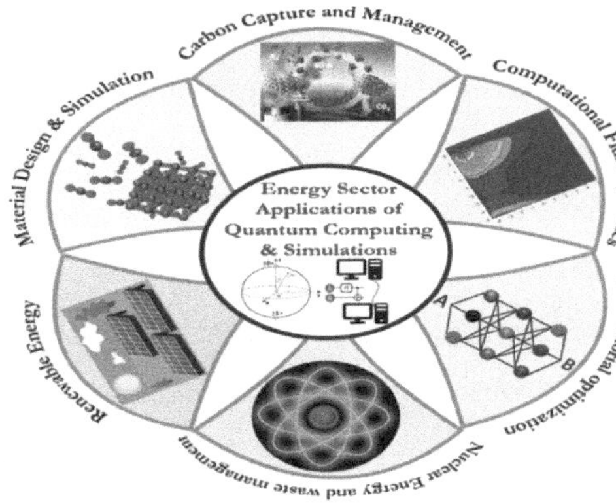

Fig 3.3 Shows role of quantum physics towards various applications

A key finding in quantum mechanics is that all energy is emitted in discrete units or bundles called "quanta." This finding provides the basis of the field's treatment of atomic and subatomic systems and their interactions. The remarkable thing is that, in most cases, the properties of subatomic particles, as seen through the lens of wave functions, can only be approximated by using quantum theory. The wave equation used to solve for wavefunctions, the Schrödinger equation serves the same purpose in quantum mechanics that Newton's laws and conservation of energy do in classical mechanics: it predicts the future behaviour of a dynamic system.

As may be seen from the line spectrum of the chemical element represented by an atom, for instance, the light or electromagnetic radiation that an atom emits or absorbs has certain frequencies (or wavelengths). The quantum theory explains that these frequencies are directly related to the energy differences between different levels of an atom's electrons, and that these differences in turn correspond to specific photon energies. When an electron moves from one level to another, it emits or absorbs a quantum of energy with a frequency that is directly proportional to the energy difference. The quantisation of light was further validated by the photoelectric effect.

3.4 Technological advancements

The study of the macroscopic and microscopic physical characteristics of matter, particularly in its solid and liquid phases, as a result of the electromagnetic interactions between electrons and atoms is known as condensed matter physics. More broadly, the topic pertains to matter in its condensed form, which consists of complex systems with several components that are strongly interdependent on one another. A few examples more unusual condensed phases are liquid crystals, ferromagnetic and antiferromagnetic phases of spins on crystal lattices of atoms, Bose-Einstein condensates in ultracold atomic systems, and the superconducting phase shown by some materials at very low cryogenic temperatures. By conducting experiments to measure different material properties and by applying mathematical models derived from theories of statistical mechanics, quantum mechanics, and electromagnetics, among others, condensed matter physicists attempt to understand the behaviour of these phases.

Fundamental forces and particles that make up matter and radiation are the focus of particle physics, often known as high-energy physics. Additionally, nuclear physics is concerned with studying combinations of protons and neutrons, but the subject as a whole analyses combinations of fundamental particles up to that size.

The Standard Model classifies the most basic particles in the universe as bosons (force-carrying particles) and fermions (matter particles). Even though ordinary matter is solely composed of first generation of fermions, there are three generations total. The proton, neutron, electron, and electron neutrino constituents of the first generation are the up and down quarks. Electromagnetism, the weak interaction, and the strong interaction are the three basic interactions that are known to be mediated by bosons.

Hadrons are necessary for quarks to exist. Baryons are hadrons with an uneven number of quarks, while mesons are those with an even number of quarks. As a whole, most matter is composed of only two baryons: the proton and the neutron. Even the most stable mesons may be destroyed in a matter of hundredths of a microsecond. The fast-moving protons and neutrons in cosmic rays, which are quark particles, collide to produce these. Cyclotrons and other particle accelerators may also make mesons.

There are identically massed antiparticles that have the opposite electric charge to a particle. The positron is an example of an antiparticle; it is opposed to the electron. Positrons have a positive electric charge, whereas electrons have a negative one. It is theoretically possible for antiparticles to combine to create antimatter. For example, the photon is both a particle and its own antiparticle.

Chapter 4 Advances in Electrochemistry and Applications

About the authors

1-Laiq Ali, a BS Chemistry graduate from Institute of Chemical Sciences Bahauddin Zakariya University, Multan, specializes in Physical chemistry. His research focuses on developing nanosensor-based electrochemical techniques for microbial detection in water. He integrates nanomaterials to enhance sensitivity and selectivity in environmental monitoring. His work in Electro and material chemistry contributes to advancing rapid and cost-effective sensing technologies.

2-Hasnain Ali From Department of biomedical engineering, University of Malaya, Malaysia. Sensor and Diagnostic Lab PUSC Lahore. Synthesis and characterization of nanomaterials, development of polymer matrix nanocomposits, eletrocatalysis, photocatalysis and electrochemical(bio) sensor.

3-Muhammad Yaseen From Department of Biochemistry, The University of Haripur. My research interest lies in the discovery of novel drugs derived from natural plant nanoparticles, with a focus on evaluating their potential anti-cancer activity and enhancing the efficacy of chemotherapeutic agents. By exploring the therapeutic properties of plant-based nanoparticles, I aim to develop innovative, less toxic alternatives to traditional chemotherapy drugs and detection of drugs as biomarkers in the body of animals using electrochemistry (oxidation and reduction reaction).

4-Dilawaiz from Department of Chemistry, University of Education Lahore, Vehari Campus. My research integrates advances in electrochemistry with the development of polymer-inorganic hybrid materials for energy storage, electrocatalysis, and CO_2 reduction. By leveraging nanofillers, biohybrids, and functional coatings, I enhance electrochemical performance, stability, and recyclability. This approach enables next-generation multifunctional materials for sustainable energy and environmental applications.

4.1 Introduction to Electrochemistry

One well-known modern field that addresses societal challenges in a number of areas, such as energy conversion and storage, processes, environmental research, and (bio)analytical chemistry, is applied electrochemistry (AE). The energy industry uses the electrochemical techniques for the energy storage and conversion. The development of effective and sustainable technologies, such as batteries, fuel cells, and electrolysers. The transition to greener and more sustainable energy choices is therefore encouraged by these advancements, which also make it easier to integrate renewable energy sources.

Innovative approaches to water purification and pollution reduction are provided by electrochemistry in environmental field. Heavy metal removal, organic pollutant degradation, and effluent biodegradability enhancement are all accomplished by electrochemical techniques, such as advanced oxidation procedures and direct and indirect electrochemical oxidation. These technologies provide effective and affordable ways to improve the water quality, preserve ecosystems, and address environmental problems.

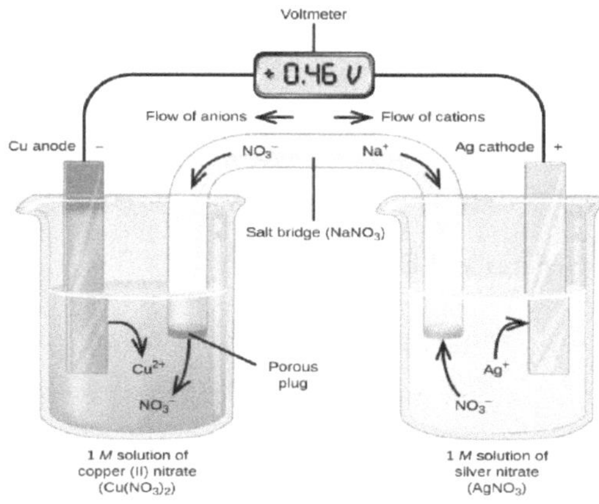

Fig 4.1 Shows the principles of electrochemistry and its emergence

For surface modification, and electrosynthesis, electrochemistry is crucial in the industrial sector. To improve the performance of materials and components, functional and protective coatings are made using electrochemical techniques such as electroplating, electroforming, and electrochemical machining. Superior products with improved material properties and remarkable durability may be produced thanks to these processes. In metallurgical industries, electrochemical surface treatments are intended mainly to produce porous materials with a larger geometric surface area and, secondarily, to aid in the development of a strong metal oxide layer that protects and stabilises the metal or semiconductor's nanostructure.

4.2 Emerging Features

Electroanalysis is also used to characterise a variety of substances and materials, perform studies, and evaluate antioxidant activity. Impedance spectroscopy, potentiometry, amperometry, and voltammetry are among the methods that these procedures. For the detection of the biomolecules, environmental contaminants, medications, and the clinical diagnostics, electrochemical and biosensors are used. These gadgets provide a number of advantages, such as affordable technologies and quick, sensitive instruments, making them indispensable equipment for quality assurance, research, and real-time process monitoring.

4.3 Energy Conversion and Buildup

Energy permeates everything and keeps us going in our pursuits. The ease of acquiring energy to run the gadgets, appliances, machines, and automobiles is something often overlook. Nevertheless, energy storage techniques must be taken into account for use. Because of their intermittent availability, renewable energy sources like wind and sunlight need an intermediate storage solution, unlike fossil fuels, which can be easily stored and transported in their unprocessed state. For energy storage, batteries are considered the only viable solution, allowing for use when needed and acting as an essential bridge between energy production and consumption.

Batteries may be divided into two categories: single-use batteries, like non-rechargeable batteries, and multi-use batteries, like rechargeable batteries. In many applications, batteries that are known for their ability to continually store and discharge energy, which leads to cost effectiveness and environmental sustainability. The various types of commercial batteries used for energy storage, along with their main applications, advantages, and disadvantages. Batteries are used in three main sectors: transportation and automotive, which includes electric vehicles (EVs); portable electronics; and stationary power storage. Each of these sectors has specific requirements.

4.3 Technological operations

The development of electric vehicle technology is heavily reliant on improving battery technology, which is beset by problems such as the poorly manufactured batteries and challenges in real-world application. Since lead-acid batteries (Pb-A) have a significant market share in terms of sales value and energy production, they have historically dominated the rechargeable battery market, particularly in the automotive sector. However, there are inherent limitations to Pb-A batteries, including a short cycle life, low energy density, susceptibility to acid stratification and leakage after damage, and challenges with miniaturisation because of problems with lead production. The environmental effects of well-established, necessitating extensive recycling programs mitigate their negative effects.

Lithium-ion batteries (LIBs) have thus emerged as a competitive substitute, gaining traction as a consequence of their many advantages, including almost 100% storage efficiency and a range of chemistries that make them suitable for a number of sustainable applications. The cells of a conventional lithium-ion battery are composed of a positive electrode called the cathode and a negative electrode called the anode. An insulating separator separates the two electrodes, and an electrolyte surrounds them. At the cathode, lithium ions attach to the active material after migrating from the anode and passing through the separator during discharge. The anode and cathode simultaneously emit and transfer electrons via an external circuit. The movements of lithium ions and electrons are reversed by a connected power source when the lithium-ion battery is being charged.

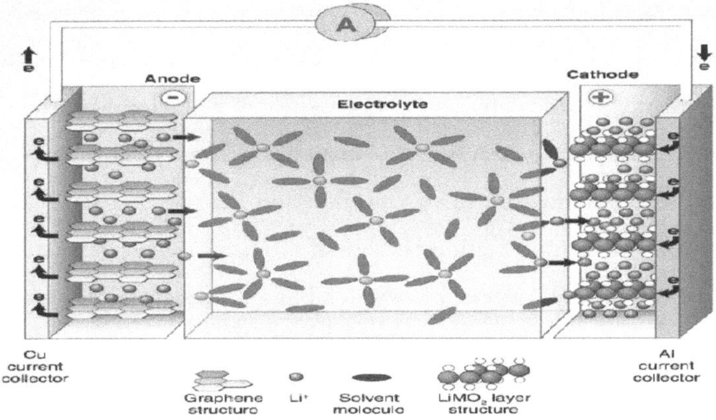

Fig 4.2 Shows the principles features of lithium-ion batteries

Anodisation, surface etching, electroplating, electrodeposition, and electrochemical functionalisation are some of the methods that fall under the umbrella of electrochemical surface modification. This section provides a concise overview of fundamentals of each of the previously listed processes and highlights the main uses for each strategy.

Sometimes terms electrode plating and electrodeposition are used interchangeably. In technical contexts, they indicate different procedures. Using an electrochemical process to put a thin metallic coating to a substrate surface is known as electroplating. It is widely used in the metallurgical industries, such as electronics, jewellery, and automobiles (e.g., nickel, platinum, gold, copper, etc.), to create protective coatings that shield surfaces from adverse conditions. This enhances material performance and improves the physical characteristics of components (lustre, form, etc.). The process of applying material to a substrate surface using electrochemical methods is known as electrodeposition, on other hand.

Electrodeposition makes it easier to produce high-quality products with improved material properties and remarkable durability. It makes it easier to electrodeposit alloys, composites, or non-metallic materials e.g., glasses, silicone, and conductive polymers like PEDOT, polypyrrole, and polyaniline. In conclusion, the application of metallic coatings is the goal of electroplating, a kind of electrodeposition.

4.4 Electrochemical Cell

A chemical reaction that happens on its own and causes a system's Gibbs free energy to decrease is called spontaneous. In electrochemistry, chemical energy is converted to electrical energy by a spontaneous process known as a redox reaction. The opposite scenario, in which electricity is used to cause a chemical reaction that is not spontaneous, is possible. An equipment called an electrochemical cell is where the interconversions take place.

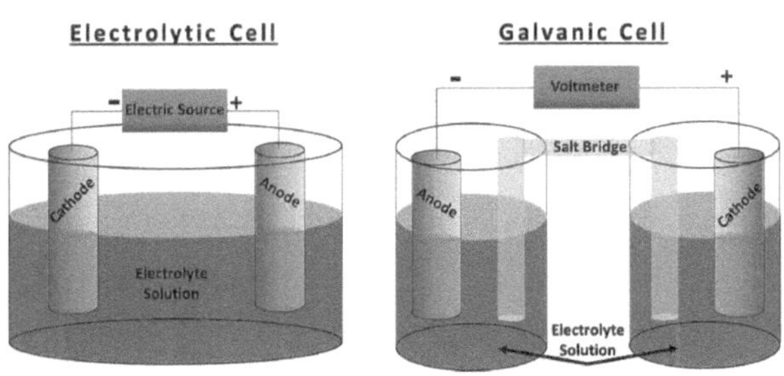

Fig 4.3 Shows the various components of electrochemical cell

An electrochemical cell is a device that uses chemical reactions to generate electrical energy. In these cells, chemical reactions may be triggered by electrical energy. Electrolytic cells are those that promote chemical processes, as via electrolysis, while voltaic or galvanic cells are electrochemical cells that generate an electric current.

Two half-cells with different oxidation and reduction processes may be thought of as being included in galvanic and electrolytic cells.

A device that uses electrical energy to drive chemical reactions or produces electrical energy from chemical processes is known as an electrochemical cell. Chemical energy may be converted into electrical energy or the other way around using these technologies. A standard 1.5-volt cell, which powers a variety of electrical gadgets such as clocks and television remote controls, is a typical example of an electrochemical cell.

Cells that have the ability to generate an electric current via internal chemical reactions are known as galvanic or voltaic cells. On the other hand, electrolytic cells are those that allow chemical reactions to occur when an electric current flows through them.

In order to create a chemical reaction that would not happen on its own, an electrolytic cell need an external electrical energy source. The cell's two electrodes—the positively charged anode and the negatively charged cathode are submerged in an electrolyte solution, and a voltage is applied across them to provide the external energy. A galvanic cell, on the other hand, produces electrical energy and is the building block of a battery. While the net reaction in a galvanic cell is spontaneous, as shown by a negative Gibbs free energy, the net reaction in an electrolytic cell is non-spontaneous (the opposite of a spontaneous reaction), as indicated by a positive Gibbs free energy.

4.5 Advancements and applications

An external voltage is used by an electrolytic cell to carry current, enabling a chemical reaction that is not spontaneous. A spontaneous chemical reaction proceeds in a galvanic cell to produce an electric current. An electrolytic cell and a galvanic cell are separated by an equilibrium electrochemical cell.

There is no current flow because an opposing electromotive force properly balances the tendency of a spontaneous reaction to generate a current through the external circuit. When this counter-electromotive force is increased, the cell becomes electrolytic; when it is decreased, the cell becomes galvanic.

Three elements make up an electrolytic cell: an electrolyte and two electrodes, known as the cathode and an anode. Typically, the electrolyte is a mixture of water or other solvents with dissolved ions. Molten salts, such as sodium chloride, may also be used as electrolytes. The ions in the electrolyte are pulled towards the electrode with the opposite charge when exposed to an external voltage between the electrodes, promoting charge-transfer activities, also referred to as faradaic or redox reactions. A normally stable or inert chemical material in solution may only be broken down by an electrolytic cell when it is exposed to an external electrical potential (also known as a voltage) of the right magnitude and polarity. It is possible for the electrical energy to cause a chemical reaction that would not occur spontaneously.

The electrode to which cations (positively charged ions, such silver ions Ag^+) travel within the cell and undergo reduction by contact with electrons (negatively charged) from that electrode is known as the cathode, according to Michael Faraday. In the same way, he defined the anode as the electrode to which anions (negatively charged ions, such chloride ions Cl^-) travel within the cell in order to undergo oxidation by moving electrons to the electrode. An external wire attached to a galvanic cell's (or battery's) electrodes creates an electric circuit, with the cathode being identified as positive and the anode as negative. Through the external circuit of a galvanic cell, positive electric current flows from the cathode to the anode.

Electrochemical cells known as galvanic or voltaic cells use spontaneous oxidation-reduction processes to produce electrical energy. When creating the equations, it is often helpful to split the oxidation-reduction processes into half-reactions in order to emphasise the particular chemical changes and help balance the equation as a whole.

Electrons are moved from one species to another during oxidation-reduction reactions. When the reaction occurs naturally, energy is released. As a result, the energy released is put to work doing good work. This energy must be addressed by splitting the reaction into two separate half-reactions, namely. reducing and oxidising. In order to promote the movement of electrons from one terminal to the other, the reactions are carried out within two separate containers using wire. It creates a voltaic cell.

The Gibbs free energy of the voltaic cell's spontaneous redox reaction provides the majority of the electrical work that a galvanic cell does. A salt bridge and two half-cells are usually included. An electrolyte submerged in a metallic electrode makes up each half-cell. Through metallic wires, the two half-cells are connected outside to a switch and a voltmeter. Sometimes there is no need for a salt bridge when both electrodes are submerged in the same electrolyte.

Electrons are retained at the electrode in a galvanic cell because the metal atoms on the electrode have a tendency to produce ions in the electrolyte solution when they come into contact with it. As a result, the metal electrode becomes negatively charged.

At the same time, metal ions in the electrolyte solution have a tendency to adhere to a metal electrode. The electrode thus gains a positive charge.

Depending on the inclinations of two opposing processes, charge separation may place under equilibrium circumstances, and the electrode may be positively or negatively charged. A potential difference between the electrode and the electrolyte is therefore created.

Electrode potential is the term used to describe this voltage. The anode, which is the electrode where oxidation takes place, and the cathode, which is the electrode where reduction takes place, are the two electrodes. Anode potential is negative with respect to the solution, whereas cathode potential is positive with respect to the solution. The result is a potential differential between the galvanic cell's two electrodes. We call this voltage the cell potential. The cell potential is also known as the galvanic cell's electromotive force when no current is drawn from it.

An external circuit, an anode, a cathode, and a load are among the fundamental parts of galvanic cells. Electrons from the related chemical species are lost as a result of oxidation reactions that take place at anodes. They are the electrodes where electron acquisition, or reduction, takes place. Electron flow between anodes and cathodes is controlled by the external circuit, and the load represents the circuit connection between the battery and the attached devices that use the electricity produced. Within each half-cell, reduction-oxidation processes are segregated, and the galvanic cell's internal electrical circuit is completed by the salt bridge. A salt bridge is not required when both half-cells are submerged in the same electrolyte; instead, a porous membrane that promotes circuit completion may be used.

Uses

Electrolysis, where "electro" means electricity and "lysis" means breaking apart, is the process by which chemical compounds are broken down using electrolytic cells, which are often used. An electrolytic cell is used to electrolyse substances like copper, silver, nickel, or chromium. Notable examples of electrolysis include the dissociation of water into hydrogen and oxygen and the breakdown of bauxite into aluminium and a number of other compounds. Direct electric current is used in the process of electrolysis.

Commercial applications for electrolytic cells include the electrorefining and electrowinning of a number of non-ferrous metals. Most industrial production of high-purity lead, copper, zinc, and aluminium occurs in electrolytic cells.

Water may undergo electrolysis, as was previously indicated, particularly when ions are added (as in salty or acidic water). A reduction process produces hydrogen gas when hydrogen (H+) ions go to the cathode and react with electrons under the influence of an external voltage source. Likewise, hydroxide (OH) ions go to the anode to release electrons and a hydrogen (H+) ion, which generates oxygen gas via oxidation.

Electrons are released to the anode when current flows through molten sodium chloride (NaCl) and oxidises chloride ions (Cl) to chlorine gas. In a similar manner, the cathode reduces sodium ions (Na+), which take up electrons from the cathode and then deposit them as sodium metal.

4.6 Chemical reactions and components

Electrolysis of sodium chloride in water is also possible. Chloride ions (Cl) are oxidised by the anode to produce chlorine gas (Cl2). Water molecules are converted to hydrogen gas (H2) and hydroxide ions (OH) at the cathode rather than sodium ions to sodium metal. An aqueous sodium hydroxide

(NaOH) solution, hydrogen gas, and chlorine gas are the products of electrolysis.

A battery is a collection of electrochemical cells that produce electricity. Every galvanic cell has the potential to be a battery in principle. A perfect battery would be continuously charged, have a steady voltage, and withstand extreme weather conditions including heat and humidity. Real batteries strike a balance between desirable qualities and practical limitations. The average car or light-duty truck's mass is around 1% that of a vehicle battery. Although this smartphone battery may provide almost infinite energy, its weight would make it inappropriate for this use. Therefore, it is impossible to identify a single battery as the "best"; instead, batteries are selected for certain uses based on criteria including weight, cost, reliability, and current size. The main and secondary batteries are the two basic types. Below is a description of a few of each kind of battery.

Batteries are the most common use of electrochemistry. One kind of voltaic electrochemical cell is the non-rechargeable battery used in television remote controls. A current of electricity flows through the electrolyte in the middle, via the negative electrode at one end, and out through the positive electrode. Energy produced by the battery's redox process is released from the battery.

This basic concept is found in almost all batteries. In modern batteries, lithium-ion cells are the most common minerals and chemicals used. New electrolyte materials including sodium, magnesium, and saltwater are being studied by scientists as a result of the growing need for batteries.

Numerous applications in sanitation make use of electrochemistry, namely the electrolysis process. Using electricity to start chemical reactions is known as electrolysis. Many companies use this process to separate brine into chlorine and sodium hydroxide.

Manufacturers use these chemicals into a variety of cleaning solutions once they have been separated. Water purification systems, disinfectants, and antiseptics are just a few of the goods that include chlorine. These substances would be more expensive or more difficult to get without electrolysis.

As with batteries, it is crucial to keep the two products produced by this kind of electrolysis apart. After being separated, they may be re-mixed to create sodium chlorate, a substance found in bleach.

Chapter 5 Advances in Biochemistry and Gene Expression

Rubab Yousaf is a M. Phil biochemistry scholar from University of agriculture Faisalabad. She is Laboratory technologist in USA. She work focuses on applying molecular techniques in clinical diagnostics and exploring recombinant DNA tools for disease detection and therapeutic monitoring.

5.1 Gene and its components

The genetic material that specifies how a protein is to be made is called a gene. The nucleus houses the DNA, the cell's information bank. The DNA code contains all the information our cells need to make proteins.

Genes store specific instructions for making a certain protein or carrying out a specific function; these instructions often encoded. The aforementioned genes undergo transcription into messenger RNA (mRNA) before being translated into a polypeptide chain. Afterwards, a polypeptide is transformed into a protein. Gene expression is the process by which all of the instructions encoded in our DNA manifest as our unique physical characteristics.

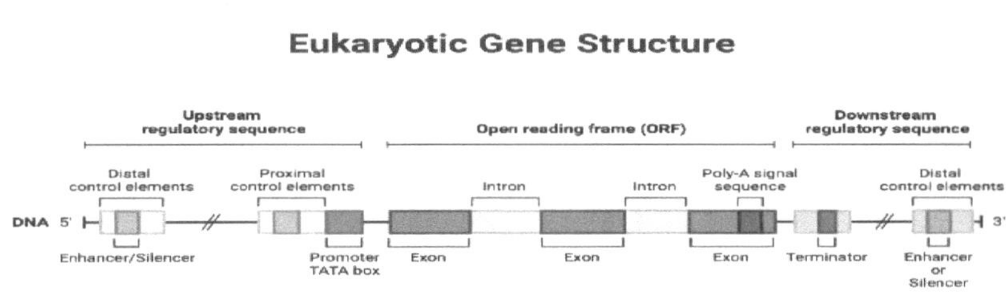

Fig 5.1 Shows the structure of gene of eukaryotic system

5.2 Role of gene expression

The translation of genetic instructions into a biologically active molecule, such a protein, is known as gene expression. A cell may adapt to its changing environment via this mechanism, which is highly coordinated.

The process of gene expression involves the translation and transcription of genetic information derived from DNA into a protein. A person's physical characteristics are a product of their genes, and this process is shown via genetic expression. This is the mechanism by which instructions are conveyed from genes to proteins.

The Keratin gene as an example to help us comprehend this issue better. The hair, nails, and skin all benefit from the protein keratin. The normal rate of growth for these is constant, as our hair, nails, and skin all gradually wear down with time. If body produces too much keratin, you may end up with a lot of hair, dry, hard skin, and long, thick nails. The only way to prevent this is to control keratin gene expression.

Gene expression regulation encompasses a wide range of cellular processes that allow us to control the quantity of protein that our genes encode.

Because prokaryotes are unicellular and do not have a nucleus, their DNA is free to move about in the cytoplasm of their cells. Transcription (from DNA to RNA) and translation (from RNA to protein) happen almost simultaneously during protein synthesis. Transcriptional regulation ends when there is no longer a necessity for the produced protein. Therefore, the main way to regulate the kind and amount of the proteins produced in prokaryotic cells is via regulation of transcription. The following procedures run without human intervention. The additional transcription happens when there's a need for the additional protein.

Thus, transcriptional regulation of gene expression is primarily responsible for prokaryotic cell biology.

The lacZ gene produces the enzyme β-galactosidase, which is responsible for converting the disaccharide lactose into Galactose and glucose. The lacY gene is responsible for encoding the membrane protein lactose permease, which helps lactose enter cells. Little is known about how the lacA gene, which codes for a transacetylase, contributes to the metabolism of lactose.

To differentiate it from structural genes, the regulatory gene I is located to the left of the lac Z gene. Regulatory genes are responsible for encoding proteins that regulate transcription by interacting with regulatory DNA sequences linked to genes. The DNA sequence that regulates transcription is the operator region that separates the I and Z genes.

Because glucose is the preferred carbon and energy source for these cells, the *E. coli* lac operon is often quiet or suppressed. When there is enough glucose, the lac operon cannot be transcribed because a suppressor protein, which is the result of the I gene, binds to the operator. When glucose levels are enough, cells will not use lactose as an alternate source of energy and carbon, regardless of its availability. On the other hand, the three enzyme products are translated when glucose levels decrease, activating the lac operon. In this section, how glucose limitation causes derepression and direct activation of the lac operon, resulting in maximum lac gene transcription only when required (i.e., when lactose is present and glucose is absent). In order to better comprehend lac operon and other aspects of *E. coli* gene regulation, some of the seminal research that paved the way for this knowledge.

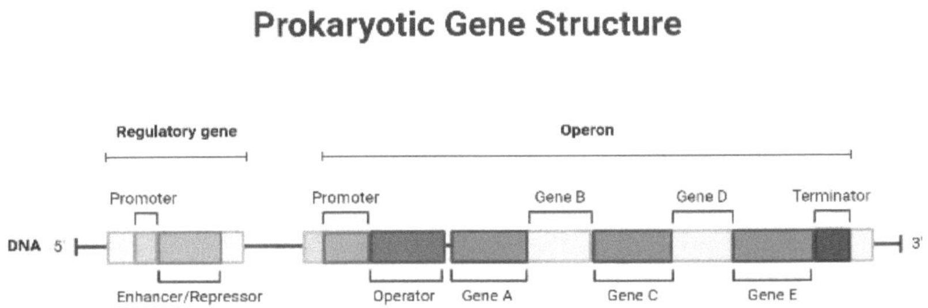

Fig 5.2 Shows the structure of gene of prokaryotic system

A second regulatory mechanism that controls lac operon expression is CAP, which stands for cAMP receptor protein or cAMP-bound catabolite activator protein. When glucose is present, CAP is in an inactive conformation and cellular cAMP levels are low. Conversely, CAP activation occurs when cAMP levels increase in response to low glucose levels. The cAMP-bound CAP will be ineffective if the lactose levels are also low. In the presence of lactose and low glucose levels, the lac repressor dissociates from the operator region due to the binding of allolactose. Here, the CAP coupled to cAMP may replace the repressor protein in binding to the operator.

Here, instead of inhibiting RNA polymerase, activated Camp-bound CAP promotes even an more effective transcription of the lac operons. The end consequence is an increase in lac enzyme production, which allows cells to utilise lactose more efficiently instead of glucose for energy. The following shows the lac operon at its most active state under conditions of low glucose and high lactose.

On the other hand, eukaryotic cells are more complicated due to presence of internal organelles. The process of converting DNA into RNA takes place inside the nucleus of eukaryotic cells. The process of RNA translation into protein begins with its export from the nucleus and entry into the cytoplasm by ribosomes.

Transcription takes place only within the nucleus, whereas translation takes place only in the cytoplasm outside of the nucleus, due to the physical separation of the two processes by the nuclear membrane. At every point in time, gene expression may be regulated. Epigenetics involves uncoiling DNA from nucleosomes so the transcription factors can bind; the transcription (transcriptional level); post-transcriptional processing and export to the cytoplasm; translation (translational level); and post-translational processing and export to the cytoplasm are all levels at which regulation can occur.

The process by which a cell determines which of the many genes included in its genome are "turned on" (expressed) is known as gene regulation. Despite sharing almost identical DNA, gene regulation ensures that virtually every cell type in your body has its own unique collection of functional genes. Each kind of cell is highly specialised for its function because of the specific patterns of gene expression that determine which proteins are present in each cell type.

5.3 Mutational events

Mutations in DNA may happen as cells divide. As the cells proliferate, they meticulously replicate the original document, which serves as the body's instruction manual. While replicating, cells may accidentally insert (add), delete (remove), or substitute (replace) letters, leaving a lot of space for mistake during cell division.

The cells may not be able to understand your genetic instruction manual if a mistake (genetic mutation), or it may be incomplete or include extraneous pieces. Because of all of this, cells may be unable to carry out their usual functions.

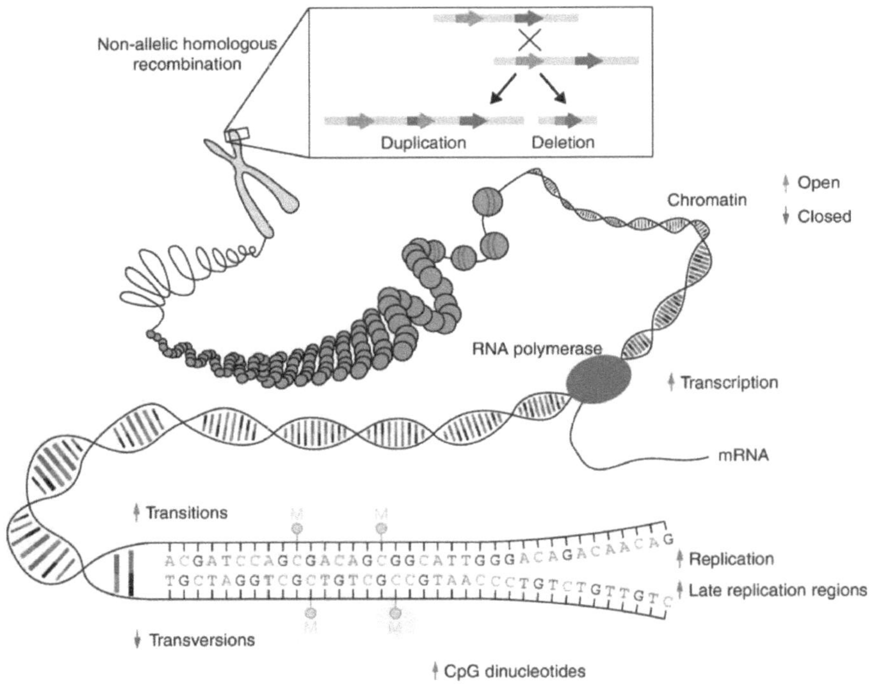

Fig 5.3 Shows development and function altered by a genetic mutation

The cells' instructions for development and function may be altered by a genetic mutation. Proteins encoded by genes instruct the body on the ideal appearance for a certain trait. Symptoms of a genetic disorder may manifest in a person with a genetic mutation because their cells are unable to carry out their normal body functions. Depending on which gene is mutated determines the symptoms of hereditary diseases. Mutations are the root cause of a wide variety of illnesses and health problems. Possible symptoms that you may encounter are:

Varieties in genes play a crucial role in evolution, the process by which species change and adapt throughout time. This mutation is very rare and affects just one individual. This person's genetic mutation is passed down across generations since it is hereditary. If the mutation increases the likelihood of survival or illness resistance, then it starts to be handed on across generations. The mutation transforms from an abnormal gene variation into an ordinary gene when it is passed down through generations of humans.

5.4 Sorts of genetic changes

Mutations may take several forms, each caused by a unique set of circumstances. Some of these are briefly summarised here:

Changes to a single nucleotide base in DNA sequence are known as point mutations. These have the potential to cause a variety of results:

When a nucleotide changes but the protein's amino acid sequence remains unchanged, this is called a silent mutation. As seen below, for instance, the identical amino acid, valine, may be coded for by switching a codon from GTA to GTT.

A missense mutation is a change in the nucleotide sequence that causes the protein to include a different amino acid. This may influence how proteins work. A missense mutation would be the change from proline (CCC) to threonine (ACC), for instance.

Nonsense Mutation: A nucleotide alteration that causes a codon to become a stop codon, resulting in the early end of protein synthesis.

If this causes termination to occur well before the typical stop codon, the protein is likely to be inactive.

This kind of mutation is known as an indel, and it occurs when one or more nucleotides are added to or removed from the DNA sequence.

A frameshift mutation occurs when the number of nucleotides added or removed is not divisible by three. This produces a change in the reading frame, which in turn affects all codons downstream. This may lead to a protein that doesn't do anything at all.

An in-frame mutation may arise when an amino acid is added or removed in increments of three, without shifting the reading frame. This can have an effect on how the protein functions.

An additional copy of a gene or chromosomal region may result via the duplication, which occurs when a section of DNA is copied. This may lead to an increase in the dosage of gene products, which might lead to illness if the disruption of normal cellular function is caused by the increased gene expression.

Inversion is the process of turning a chromosomal section upside down. An inversion involves turning the afflicted section upside down, which may interfere with the genes' ability to work there. The centromere is involved in pericentric inversions, but paracentric inversions do not.

The loss or deletion of a chromosomal segment results in the elimination of crucial genetic information. This may have the devastating effects on the development or function, depending on the genes impacted and the magnitude of the deletion.

One kind of chromosomal mutation is known as a translocation, which may happen within a single chromosome when one segment of the chromosome moves to another.

A chromosomal anomaly known as translocation occurs when sections of two distinct chromosomes are swapped out. Two primary varieties exist: a reciprocal A chromosomal translocation occurs when two chromosomes that are not identical swap genetic material. When this happens inside a gene or a region that regulates it, it might interfere with gene function. Two acrocentric chromosomes fuse to form a single chromosome in a Robertsonian translocation. Genetic diseases like Down syndrome often include this kind of translocation.

The process of insertion (interchromosomal) involves the transfer of genetic material from one chromosome to another. If the insertion happens in a crucial gene or regulatory area, it might affect gene function.

Chromosome fusion is the process by which two smaller chromosomes join to form a bigger one. Gene expression and regulation may be impacted by this.

Expansions of Repeat Sequences: The Huntington gene, for example, has CAG repeats, which are repeating sequences of nucleotides. The amount of these repeat units increases with expansion mutations. There is a cutoff point beyond which the amount of repetitions starts to cause illness. An overabundance of CAG repeats, for instance, causes neurodegeneration and aberrant protein aggregation in Huntington's disease.

Chapter 6 Advances in Biotechnology and Plant Physiology

Introduction to Authors

1-Tahir Yab Khan is MS.c(Hons) Scholar in Agriculture - Horticulture from Ghazi University Dera Ghazi Khan Pakistan. Master Research: Varietal performance of seedless watermelon (*Citrullus lanatus* L.) accessions under Agro-Ecological conditions of Layyah, Pakistan. Research Interest: Environmental biology and Plant stress physiology to overcome the impact of environmental stresses on the growth and productivity of crops by improving the genetic make of plants.

2-Rabbia Akhtar working as a M. Phil Scholar at Department of Botany, University of Agriculture, Faisalabad. Research interests: Plant Physiology (specifically drought and heat stress), ethnobotany and biochemical activities of plants. Mainly how plants cope up in different environmental conditions and how their anatomical and morphological features adapt such stresses.

6.1 Introduction to plant functions

The study of how plants work is known as plant physiology, a branch of botany. To begin, plant physiology include the study of phytochemistry, often known as plant chemistry. Plants produce a vast variety of chemical substances that no other living thing requires for basic functions or survival. Many different chemicals, including pigments and enzymes, are necessary for photosynthesis to take place. Due to their immobile nature, plants also rely on chemical defenses to ward off herbivores, diseases, and other plants.

Toxins and chemicals with an unpleasant smell or taste are what they use to do this. In addition to attracting pollinators and animals to disperse mature seeds, other substances protect plants from disease, allow them to survive drought, and get them ready for dormancy.

6.2 Technological advancement in plants

The interplay between a plant's cells, tissues, and organs is the third main focus of plant physiology. To carry out their unique roles, cells and tissues undergo chemical and physical specialization. In order for a plant to stay there and get nutrients from the ground, it needs roots and rhizomes. In order for the roots and the leaves to continue functioning, there must be a continuous flow of minerals from the roots to the leaves and nutrients from the leaves back to the roots. Plants have evolved several pathways for this kind of transport, including vascular tissue, and plant physiologists study how these pathways work.

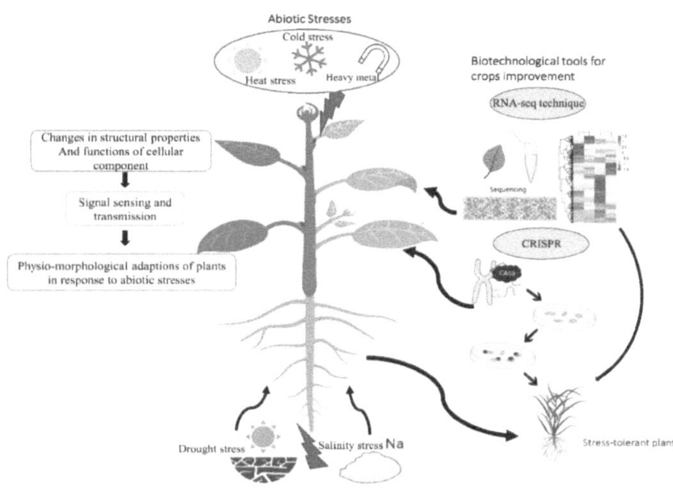

Fig 6.1 Shows the function and technological advancement in plants

As a fourth point, plant physiologists investigate how plants manage their own internal processes. Hormones are chemical messengers that plants use to instruct other parts of the plant to do certain actions, much as in mammals. A process called photoperiodism allows many blooming plants to bloom at certain times based on the duration of the night. The production of the gas ethylene by plants helps regulate the ripening of fruit and the loss of leaves throughout winter.

Lastly, environmental physiology is a subfield of plant physiology that studies how plants react to different types of environmental stress. A plant's behavior might vary in response to stress caused by things like water loss, changes in air chemistry, or being crowded out of its space. Factors including genetics, chemistry, and physics may influence these alterations.

Pigments are some of the most crucial chemicals for plants to carry out their functions. Porphyrins, carotenoids, and anthocyanins are only a few of the many types of chemicals found in plants that serve as colors. Pigments in living things only take in light of certain wavelengths, while reflecting the rest. The plant may utilize the absorbed light to fuel chemical processes, and the color of the pigment is determined by the wavelengths of light that reflect off-of it.

6.3 Various functional aspects in plants

Plants rely on the chlorophyll, a porphyrin that filters out blue and red light while retaining green light, as their principal pigment. Plants get their characteristic verdant hue from chlorophyll, which is present in relatively high concentrations in plants. Chlorophyll a and chlorophyll b are the two types of this pigment that are present in all terrestrial plants and green algae. Red algae have the chlorophyll a type, whereas the kelps, diatoms, and other photosynthetic heterokonts have chlorophyll c. The main way plants utilize light to power photosynthesis is via chlorophylls.

Red, orange, or yellow forms tetraterpenoids are the carotenoids. They aid photosynthesis by acting as accessory pigments in plants and absorbing light at wavelengths that chlorophyll struggles to handle. Carotene, lutein, and lycopene are the most well-known carotenoids. Carrots and onions contain carotene, an orange pigment; tomatoes include lycopene, a red pigment. Studies in humans have shown that carotenoids support good vision and function as antioxidants.

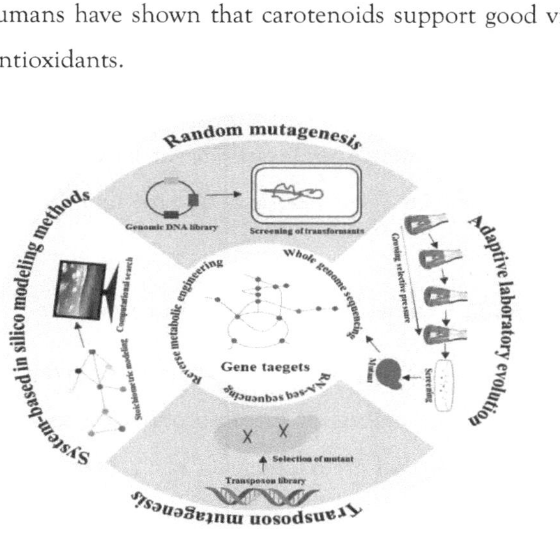

Fig 6.2 Shows emerging technologies and genetic manipulation in plants

The water-soluble flavonoid pigments known as anthocyanins literally mean "flower blue" and may take on a reddish-blue hue depending on the pH level. They are present in all parts of higher plants and provide color to things like flowers, fruits, leaves, stems, and roots, albeit often in little enough amounts that they aren't evident. Flowers' petals are a great place to see anthocyanins, which may comprise up to 30% of the tissue's dry weight.

Tropical shade plants, like Tradescantia zebrina, get their purple underside color from them as well. To make the most of the light that reaches these plants, anthocyanin absorbs it after it has gone through the leaf and reflects it back to the parts of the plant that contain chlorophyll.

Betalains may be either yellow or red in color. They are water-soluble and produced from indole, but they are not anthocyanins; instead, they are molecules formed from tyrosine. Cacti and amaranth are the only plants in the Caryophyllales family that have this kind of pigment; anthocyanins and this pigment class never coexist in plants. Beets' rich red color is accomplished via the betalains, which are also used the commercially as food coloring additives. Although the exact role of betalains in plants is still unclear to plant physiologists, there is some early indication that they may have fungicidal capabilities.

6.4 Principles illustrations of some mechanism aspects

The proper nourishment of plants is a key component in their development. Deficit symptoms, the process of the absorption, the techniques used to determine which elements are required for plant growth, the functions of these elements, and the criteria used to determine which elements are essential are all covered in this article. The significance of nitrogen fixing is also addressed.

Plants rely on both macro and micronutrients to manage vital functions including enzyme activity, the osmotic concentration of cell sap and its maintenance, cell membrane permeability, and many more.

The plant may reach critical concentration if it does not get enough of certain nutrients. A process involving distinct organs, tissues, and cells is responsible for element absorption.

As auxin builds up in the apical meristem, it triggers the process of new leaf production. In order to prevent new leaves from growing too near to existing ones, cells in the immediate vicinity of budding leaves deplete auxin. The pattern of the plant's leaves is formed in this manner. The exact patterning of the developing leaf's epidermal cells is likewise regulated by auxin.

The goal of plant biotechnology is to modify plants to meet human demands or take advantage of unique environmental conditions. There are often times when a number of needs and possibilities come together.

For instance, it's possible that only one crop is needed to provide a steady supply of the nutritious food, save the environment, create employment possibilities, and generate money. It is usually a very difficult task to find or create plants that are suited.

Transgenic (genetically altered) crops, marker-assisted selection (MAS), genetics and genomics, and other plant biotechnologies help generate novel varieties and features. Genome mapping, gene detection, gene function discovery, gene selection in breeding and genetic resources, and gene transfer into plants for desired characteristics are all made possible by these biotechnologies. In order to create and implement biotechnologies in the food and agricultural sectors, supports research, education, and extension.

6.5 Biology in a synthetic environment

In this next stage of plant breeding, plant biotechnology is seen as the game-changing technology that may assist in meeting this issue. Techniques in genomics, molecular breeding, transgenic technology, cell and tissue manipulation, marker-assisted selection, and other plant biotechnologies help produce novel plant varieties and specific features within current ones.

Crop genetic variability can be created, preserved, and used in various ways through cell and tissue culture technologies. These include: in vitro pollination, embryo rescue for distant hybridization, polyploid breeding, in vitro mutagenesis, soma clonal variation, in vitro selection, germplasm preservation (in vitro for medium-term and cryopreservation for long-term), protoplast fusion for somatic hybrid production, gene manipulation for transgenic crops, and the recently developed methods for producing gene-edited plants.

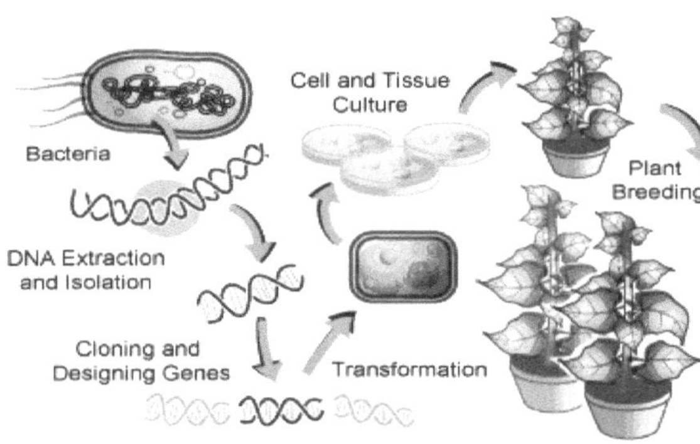

Fig 6.3 Shows the various features in the gene transfer in plants

Adapted cultivars could cross-bred with dwarfing genes during the 'Green Revolution,' leading to dramatically increased rice and wheat yields. The 'Green Revolution,' which relies heavily on technology and requires a lot of input, has proven successful in feeding the world's expanding population in the last several decades. In poor countries in particular, it guaranteed food security. But today we can see the long-term effects: soil degradation, lower groundwater levels, polluted and salinized bodies of water, and less biodiversity. Increasing the use of fertilizer is also necessary to maintain good crop yields.

When choosing for field-observable morphological features like height, grain size, color, leaf form, etc., traditional crossbreeding is simple. The "Green Revolution" primarily altered rice and wheat via dwarfing, which improved fertilizer response and decreased lodging while increasing photosynthate partitioning in grains.

Therefore, dwarf plants in the groups that were segregating could be easily identified. However, in the field, breeders face numerous factors, making it difficult to visually select for characteristics like nutritional quality, disease resistance, and abiotic stress tolerance while segregating populations. Hence, methods that can be carried out in a laboratory are necessary for the purposes of increasing genetic variety or for the purpose of genetically altering and selecting favorable genotypes.

One of Australia's most important exports is chickpeas, a kind of pulse known as *Cicer arietinum*. Globally, chickpeas rank second in consumption, only behind dried beans. Drought tolerance in crops is a major focus in breeding efforts since Australia is the second driest continent in the world, just behind Antarctica. Australia and India, the world's two biggest producers of the chickpeas, follow a similar pattern: they plant the seeds in the latter days of the summer, then let them develop on soil that is gradually drying up until they reach maturity, all while the experiencing terminal drought.

Root biomass and a few morphological and physiological aspects of leaves are among the several characteristics linked to drought resistance. Crops like chickpeas, which are grown on residual moisture, might avoid dryness by maturing early. A quantitative trait loci (QTL) hotspot area for drought tolerance in chickpea has been discovered, and there is intense continuing effort to develop molecular markers for marker-assisted selection of this trait.

Through a backcrossing experiment that introduced drought tolerance from accession ICC4958 to a high yielding Indian cultivar 'JG11,' using this hotspot as a selection marker, the drought-tolerant variety 'Geletu'. But there are other, more creative ways to screen populations for drought tolerance early on in their development that might speed up breeding even more. It is found metabolites in early chickpea seedlings that may be utilized for prognostic purposes to forecast the amount of seeds produced by mature plants under terminal drought. Under abiotic stress, especially drought, the quantity of seeds, rather than their size (weight), is what is lost in annual crops in terms of yield components.

Therefore, it would be helpful to select drought-tolerant segregants early in breeding populations if metabolites could be identified that indicate seed number under the terminal drought later in the life cycle. Under these circumstances, it is possible to forecast high or low seed counts in chickpea using pinitol, sucrose (which has a negative association with seed number), and gamma-aminobutyric acid (which has a positive correlation). This is the first of its kind: a predictive marker for screening drought tolerance. Breeders will now be able to find genotypes that do well in challenging environments without actually exposing them to drought.

The most terrible cassava disease in Africa and the Indian subcontinent is cassava mosaic disease (CMD), which is caused by a collection of eight geminiviruses and spreads via the white fly *Bemisia tabaci* and contaminated planting material.

It is the most devastating plant viral disease globally, causing an estimated yearly economic loss of USD1.9-2.7 billion. Millions of subsistence farmers in Eastern and Central Africa are facing a grave danger from the recently emerging cassava brown streak disease (CBSD), which is caused by two types of ipomoviruses: Cassava brown streak virus (CBSV) and Ugandan cassava brown streak virus (UCBSV). Both viruses are spread by white fly.

By using hairpin probes that include sequences from many viruses, RNAi-based technology may be used for the concurrent control of numerous viruses. This method to create transgenic plants of the well-known African variety 'TME204,' which produce an inverted repeat construct made up of coat protein sequences from CBSV and UCBSV fused together. The resultant transgenic plants were highly resistant to both viruses, and they maintained the beneficial agronomic traits of the 'TME204' cultivar, which is popular among farmers. By 2020, CBSD-tolerant genetically modified cassava has been green-lit for planting.

Even if the cultivar's safety and disease resistance have been evaluated, there is still a lot of doubt and criticism around the release. To create mutants that can withstand CMD and CBSD, scientists have also tried to induce mutations using methods other than genetic modification. Experiments in several agro-ecological zones showed that the three mutants outperformed their parents in terms of disease resistance. However, despite a 15-20-fold increase in carotenoids, an effort to boost β-carotene concentration in cassava by the co-expressing transgenes for deoxy-d-xylulose-5-phosphate synthase and bacterial phytoene synthase led to decreased dry matter and starch content.

Chapter 7 Advances in Chemistry

Abdul Raheem From Department of Chemistry, University of education Lahore, Dera Ghazi Khan campus. I have completed MSc Chemistry (organic) from University of Education, Lahore, sub campus Dera Ghazi Khan. I am fascinated by the transformation of simple starting materials into complex, functional molecules that hold value in pharmaceuticals, materials science and sustainable chemistry.

7.1 Introduction and emerging features of Chemistry

The study of the matter and its constituent compounds is the focus of the scientific subdiscipline of chemistry. The characteristics of these chemicals and the processes they undergo to create new substances are also covered. Atoms, ions, and molecules which together comprise the elements and compounds are the main subjects of chemistry.

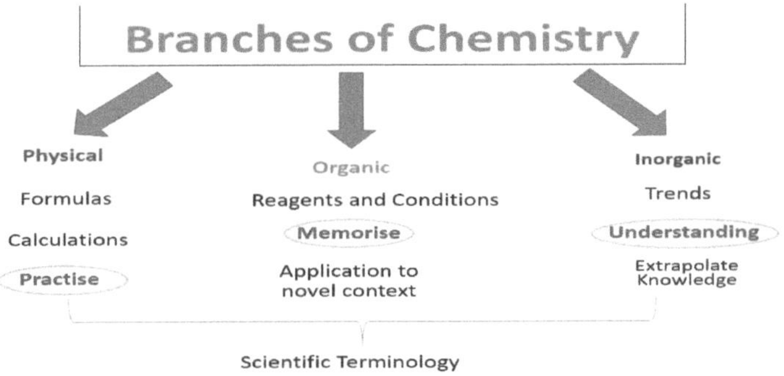

Fig 7.1 Shows Chemistry and advancement for the research development

7.2 Associations of different domains

Chemistry is the study of the characteristics, compositions, and structures of elements and compounds, as well as how they may change and the energy generated or absorbed during such changes. Chemical bonds are often the means by which these chemical species interact with one another. It is essential to remember that chemistry also studies the relationships between matter and energy.

The scientific study of the structure, characteristics, and interactions of organic compounds and materials that is, matter in all of its forms that include carbon atoms is the focus of the subdiscipline of chemistry known as "organic chemistry." The structural formula of an organic substance is determined by its structure. Physical and chemical characteristics are studied, and chemical reactivity is assessed to comprehend behaviour. The chemical synthesis of natural products, medications, and polymers, as well as the laboratory and theoretical (in silico) investigation of individual organic molecules, are all included in the study of organic processes.

Most organic substances melt, and many boil. On the other hand, inorganic materials prefer to decompose rather than boil, even if they may usually be melted. The purity and identification of organic compounds were formerly determined in large part by the melting and boiling points. The polarity and molecular weight of the molecules are correlated with their melting and boiling points. Certain chemical compounds sublimate, particularly those that are symmetrical. The odorous component in contemporary mothballs, para-dichlorobenzene, is a well-known example of a sublimable organic compound. Although there are rare exceptions, organic molecules are often not very stable at temperatures higher than 300 °C.

Solubility

Generally speaking, neutral organic substances are hydrophobic, meaning they dissolve less readily in organic solvents than in water. Organic molecules with ionisable groups and low molecular weight alcohols, amines, and carboxylic acids where hydrogen bonding takes place are exceptions. Otherwise, organic solvents have a tendency to dissolve organic molecules. There are significant differences in solubility between organic solvents and solutes.

7.3 Properties of solid states

Depending on the application, different specialised properties of organic polymers and molecular crystals with conjugated systems are of interest. For example, electro-optical (such as non-linear optics), electrical conductivity (see conductive polymers and organic semiconductors), and thermo-mechanical and electro-mechanical properties like piezoelectricity. These characteristics are primarily focus of materials science and polymer research due to historical reasons.

In organic chemistry, the idea of functional groups is essential for both classifying structures and forecasting their characteristics. The reactivity of a functional group, which is a molecular module, is thought to remain constant throughout a range of molecules, within certain bounds. The physical and chemical characteristics of organic molecules may be significantly impacted by functional groups. The functional groups of molecules determine their classification. For example, the component C-O-H is present in all alcohols. All alcohols often form esters, have a tendency to be rather hydrophilic, and can typically be transformed into their corresponding halides.

Heteroatoms—atoms other than the C and H—are present in the majority of functional groupings. Alcohols, carboxylic acids, amines, and other functional groups are used to categorise organic substances. Because of their electrical impact on neighbouring molecules, functional groups change a molecule's acidity or basicity.

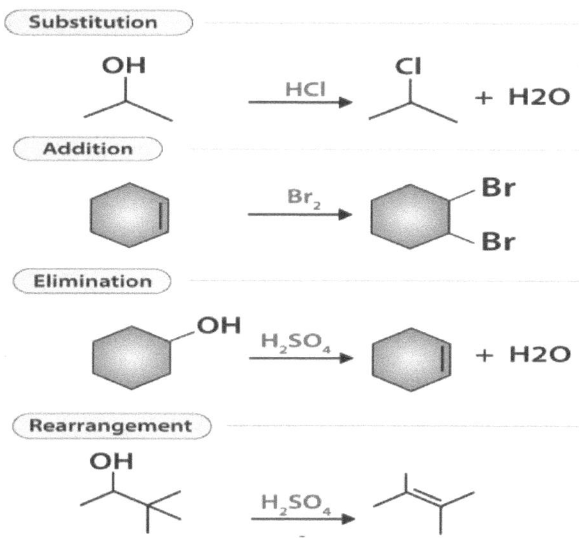

Fig 7.2 Shows the various reactions in the Chemistry

The presence of heteroatoms, which may be either endocyclic (a member of the ring itself) or exocyclic (substitutes bonded externally to the ring), changes the properties of the cyclic hydrocarbons once again. The ring is referred to as a heterocycle in the later instance. Piperidine and tetrahydrofuran are the analogous alicyclic heterocycles, while pyridine and furan are instances of aromatic heterocycles. Heterocyclic compounds often include one of three heteroatoms: nitrogen, sulphur, or oxygen, with the latter being most prevalent in biological systems.

Additionally, chemometrics, experimental design, and the development of novel measuring instruments are the main areas of attention for analytical chemistry. There are several uses for analytical chemistry in research, engineering, and medicine.

Since the beginning of chemistry, analytical chemistry has played a significant role in identifying the elements and molecules that are present in a given item. Justus von Liebig's creation of systematic elemental analysis and systematised organic analysis based on the particular reactions of functional groups were two important contributions to analytical chemistry during this time.

Instrumental analysis is the mainstay of contemporary analytical chemistry. A lot of analytical chemists concentrate on only one kind of equipment. Scholars often concentrate on either new techniques of analysis or new applications and findings. For example, an analytical chemist may be engaged in the identification of a chemical in blood that raises the risk of cancer. Using a tunable laser to improve the sensitivity and specificity of a spectrometric approach might be an attempt to create a new technique. Once created, many procedures are intentionally maintained static to allow for long-term data comparisons.

This is especially true for forensic, environmental, and industrial quality assurance (QA) applications. In the pharmaceutical sector, analytical chemistry is becoming more and more significant. In addition to quality assurance, it is used in the development of novel drug candidates and in clinical settings where it is essential to comprehend how the medicine interacts with the patient.

In more specific situations, spectroscopy is the exact study of colour as it is generalised from visible light to all bands of the electromagnetic spectrum. Spectroscopy is the branch of study that measures and analyses electromagnetic spectra.

In the domains of astronomy, chemistry, materials science, and physics, spectroscopy—mainly in the electromagnetic spectrum—is an essential investigative tool that enables the examination of the composition, physical structure, and electronic structure of matter at the atomic, molecular, and macro scales as well as over astronomical distances.

7.4 Current challenges

With components like CD/DVD diffraction gratings, smartphones, and 3D-printed parts, spectroscopy has become a popular practice within the maker movement, allowing educators and hobbyists to build the working spectrometers using easily accessible materials. These devices provide a practical way to learn about the interactions between light and matter. Open-source tools and smartphone applications make integration easier and significantly streamline the collection and analysis of spectral data. DIY spectroscopy offers worthwhile educational experiences and supports citizen scientific programs, promoting accessibility to spectroscopic methods, despite limits in resolution, calibration accuracy, and stray light control as compared to professional equipment.

The analytical measuring technique known as resonance-ionization spectroscopy (RIS) is very sensitive and selective. It splits the neutral species into a positive ion and a free electron with a negative charge by using lasers to expel electrons from certain kinds of atoms or molecules. To identify elements or compounds and ascertain their quantity in a sample, those ions or electrons are then found and counted using a variety of techniques. Since its inception in the 1970s, the RIS technique has been used in an increasing variety of fields to further our understanding of biology, chemistry, and physics. Because it provides the combined benefits of great selectivity between various atom types and sensitivity at the one-atom level, it is used in a broad range of real-world measurement systems.

Physical and chemical analyses of certain atom populations are among the uses for a basic atom counter. Different types of mass spectrometers, which provide the extra characteristic of isotope selectivity, are incorporated into more sophisticated systems. For example, these more complex RIS systems can be used to date lunar materials and meteorites, investigate ancient groundwater and ice caps, measure the Sun's neutrino output, identify trace elements in electronic-grade materials, look for resources like gold, platinum, and oil, investigate the function of trace elements in biology and medicine, identify DNA structure, and solve a variety of environmental issues.

The extremely inventive field of organic chemistry enables researchers to design and investigate molecules and compounds. The majority of an organic chemist's time is devoted to creating new chemicals and improving the synthesis of preexisting ones. Drugs used as human or animal medicines are created, manufactured, and marketed by the pharmaceutical business. Some pharmaceutical companies deal in prescription drugs and medical devices (agents that act on diseases without chemical interaction with the body) that are either brand-name (i.e., have a trade name and can be produced and sold only by the company holding the patent) or generic (i.e., a lower-cost, chemically equivalent version of a brand-name drug).

Organic chemistry is used in the majority of our consumer goods. Consider the cosmetics sector as an example. Chemists create products based on their understanding of organic chemistry, which studies how the skin reacts to environmental and metabolic stimuli.

7.5 Chromatography and its features

Chromatography is a laboratory method used in chemical analysis to separate a mixture into its constituent parts. The mixture is carried through a device (a column, capillary tube, plate, or sheet) that has a substance known as the stationary phase fixed to it after being dissolved in a fluid solvent (gas or liquid) known as the mobile phase.

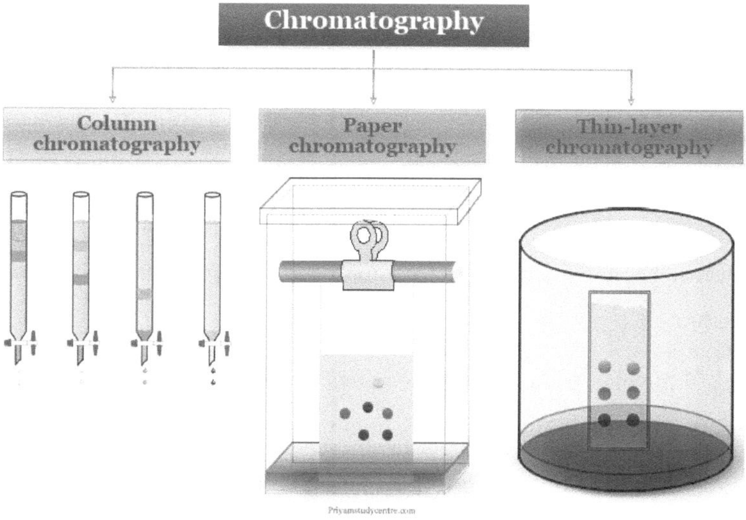

Fig 7.3 Shows the various steps in the chromatography

The mixture's constituents separate because they move through the mobile fluid at varying apparent velocities due to their varying affinities for the stationary phase and the length of time they are retained based on their interactions with its surface sites. The differential partitioning between the stationary and mobile phases serves as the basis for the separation. The separation is impacted by subtle variations in a compound's partition coefficient because they cause variable retention on the stationary phase.

Chromatography may analytical preparatory. Preparatory chromatography is a type of purification that is used to separate the components of a mixture for later use. Because of its mode of production, this process is associated with higher costs. Analytical chromatography is typically performed with smaller amounts of material and is used to determine the presence or measure the relative proportions of analytes in a mixture. There is no reciprocal exclusion between the two categories.

The foundation of affinity chromatography is the selective non-covalent interaction of certain molecules with an analyte. It is often used in biochemistry to purify proteins coupled to tags since it is very specific but not particularly robust. These fusion proteins are identified by substances that bind selectively to the stationary phase, such as His-tags, biotin, or antigens. These markers are often eliminated during purification, yielding pure protein.

The affinity of a biomolecule for a metal's cations (Zn, Cu, Fe, etc.) is often used in affinity chromatography. Columns may be made especially for the proteins of interest and are often made by hand. Conventional affinity columns are used as a first step in the analysis of a protein with unknown physical characteristics or as a preparatory step to remove undesirable biomolecules.

Nonetheless, there are liquid chromatography methods that make use of affinity chromatography characteristics. The aforementioned compounds may be separated according to their respective affinity for the metal using immobilised metal affinity chromatography (IMAC). To produce a column with a desired affinity, these columns may often be loaded with various metals.

Size-exclusion chromatography (SEC) separates molecules based on their size (or more precisely, their hydrodynamic diameter or hydrodynamic volume). It is sometimes referred to as gel permeation chromatography (GPC) or gel filtration chromatography. Smaller molecules may pass through the media's pores, trapping them and removing them from the mobile phase's flow. The effective size of the analyte molecules determines the average residence duration in the pores. Nevertheless, molecules bigger than the packing's average pore size are excluded and, as a result, experience almost little retention; these species are the first to elute. It is often saved for the last, "polishing" stage of a purification as it is a low-resolution chromatography technology. Because it can be performed in native solution, it is very helpful for figuring out the tertiary and quaternary structures of purified proteins.

A practical and efficient method for directly capturing proteins from an unclarified crude sample is expanded-bed adsorption chromatography. The equilibration buffer initially expands settled bed in EBA chromatography. After that, the crude feed—a concoction of soluble proteins, impurities, cells, and cell debris—is transported upward via the enlarged bed. Particles and pollutants go through the adsorbent, but target proteins are trapped there. The target protein desorbs in expanded-bed mode when the elution buffer is changed while the upward flow is maintained. On the other hand, if the flow is reversed, the adsorbed particles will settle fast, allowing an elution buffer to desorbed the proteins.

The properties of the feed determine whether expanded-bed or settled-bed elution is used. Following elution, a predetermined cleaning-in-place (CIP) solution is used to clean the adsorbent. Cleaning is followed by either storage or column regeneration (for future use).

Generally speaking, Hydrophobic Interaction Chromatography (HIC) works well if the sample is sensitive to harsh solvents or pH changes that are often utilised in other chromatography techniques but not excessive salt concentrations. It is often the buffer's salt content that varies. The four different kinds of hydrophobic resin in conjunction with bovine serum albumin (BSA) to illustrate how temperature affects HIC. In order to affect the binding affinity of BSA onto the matrix, the research changed the temperature. It was determined that cycling the temperature between 40 and 10 degrees Celsius would not be sufficient to completely remove all of the BSA from the matrix, but it might be very effective if the column were only used a few times. By using temperature to effect change, labs can save money by not having to purchase as much salt.

Chapter 8 Emerging Trends in Medicinal Plants and biological activities

Syda Zahra Haider from Department of Botany, Ghazi University, Dera Ghazi Khan, Pakistan. My research interests lie at the intersection of plant stress physiology, ecology, and genetics. I am fascinated by the complex interactions between plants and their environment, particularly under stressful conditions. My goal is to understand the physiological, genetic, and molecular mechanisms that enable plants to adapt to environmental stresses, such as drought, heat, and salinity.

8.1 Introduction to Medicinal Plants

For ages, people have turned to herbs and medicinal plants for their many physiologically active components. Generations of indigenous people have relied on the plant's raw materials and pure chemicals to heal a wide range of illnesses. While there is a lot of interest in studying them right now, there are several obstacles that need to be overcome during solvent extraction if they are to be used in phytochemical and biological studies. The unparalleled abundance of chemical diversity in natural goods opens almost boundless prospects for innovative drug discovery.

Between 1960 and 1997, the chance of remission for a kid with leukaemia increased by 85 percent, thanks to two medications made from the alkaloids of the pink periwinkle (Catharanthus roseus) plant in Madagascar. A new chemical found in a Madagascar plant, for example, may lead to the development of new medicines and the slowing of the spread of infections that are resistant to these drugs. Because of their unique properties as a source of medicinal phytochemicals, as well as their effectiveness, safety, and low side effects, natural products are today of great relevance. Many plant chemicals have bioactive properties. These include terpenoids, alkaloids, coumarins, flavonoids, organosulfur the compounds, phenolics, nitrogen-containing compounds, and many more. In addition to their antioxidant, antibacterial, anti-inflammatory, and immunostimulatory properties, these chemicals display a broad array of bioactivities. Because plants contain beneficial phytochemicals and the pharmaceutical and cosmeceutical industries are moving towards using natural goods, research on medicinal plants is just as vital as research on traditional medications. Chemical structures of a few crucial.

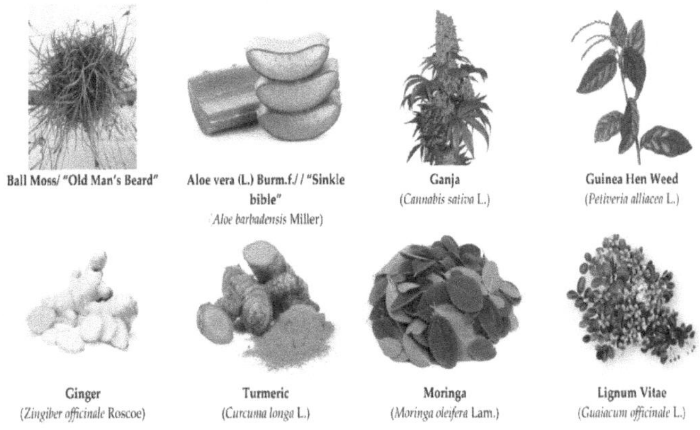

Fig 8.1 Shows the types of different medicinal plants

8.2 Bioactive compounds in Medicinal Plants

Fruits, seeds, and vegetables contain polyphenols, which are secondary metabolites, or no nutritional natural compounds, of plants. The vast family of bioactive metabolites known as polyphenols originates in secondary metabolism and is present in many plants.

The phenylpropanoid pathway is the primary route for the synthesis of polyphenols from l-phenylalanine. Polyphenols are defined by the presence of phenolic groups that may take many different structural forms; the most common of these include tannins, coumarins, stilbenes, lignans, flavonoids, and phenolic acids. Everyday foods including nuts, vegetables, fruits, and seeds contain polyphenols, which are also responsible for their colour, fragrance, and antioxidant capabilities.

Because of their positive effects on health, polyphenols are gaining prominence. Their impact as an antioxidant in the fight against oxidative stress, cancer, heart disease, inflammation, neurological disorders, and illnesses associated with ageing is growing. Pharmaceuticals, dietary supplements, and cosmetic additives are just a few of their many potential applications. There have been over 8,000 polyphenols identified in plants, with hundreds more described in cereals, veggies, and fruits.

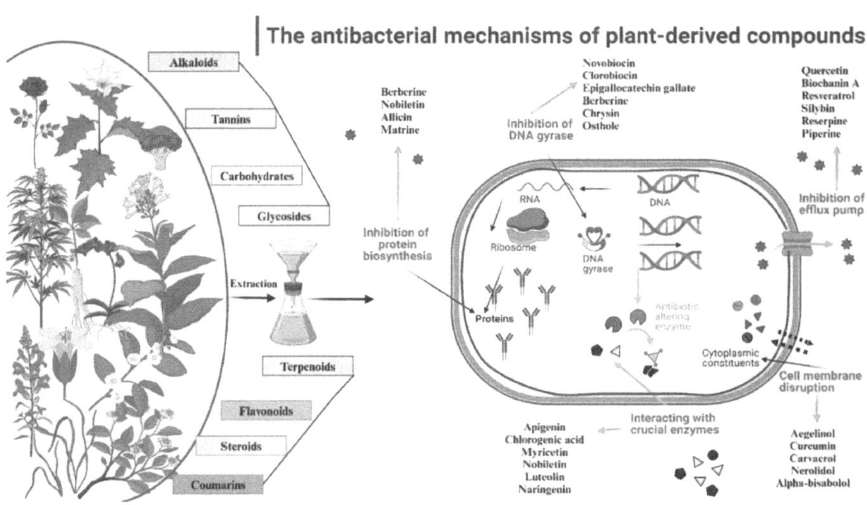

Fig 8.2 Shows the mechanism of action of compound in medicinal plants

An important function that polyphenols play is in reducing the occurrence of some malignancies by preventing carcinogenesis. They may be employed as chemopreventive drugs because of their extensive medical uses and decreased toxicity. The activation of carcinogenesis can be facilitated by inflammatory cytokines, tumour promoters, or endogenous or exogenous carcinogens through the regulation of transcription factors, proapoptotic proteins, protein kinases, growth factor signalling pathways, or cell cycle proteins. Through the elimination of reactive oxygen species (ROS) and the induction of phase II conjugating enzymes, one may potentially prevent cancer by preventing carcinogens from damaging cellular DNA during the initiation phase.

Modulating apoptosis and cell cycle arrest are two important strategies for inhibiting cell growth in the following step. It is crucial to stop angiogenesis and malignant cell development to metastasis and invasiveness in the progression. It has been shown that dietary polyphenols have a protective effect on a variety of cancer-causing processes, including phase I and phase II enzymes, cell cycle progression, apoptosis, DNA repair, metastasis, angiogenesis, and cell proliferation.

8.3 Biological Activities

Free radicals are a byproduct of oxidative stress, and antioxidants are molecules with this property. Overproduction of free radicals, the culprits behind oxidative stress, is a result of modern lifestyles, but all living things include inherent defence mechanisms that scavenge free radicals. The cellular components, such as DNA, lipids, and proteins, were all subject to oxidative stress's destructive effects. The treatment of oxidative stress has been the subject of several pharmacological medications.

The carcinogenicity and high cost of synthetic antioxidants are two of their major drawbacks. The natural antioxidants such as vitamin C, vitamin E, xanthophylls, tannins, carotenes, and phenolic have multiple biological activities. These have been investigating these compounds as potential alternatives to synthetic antioxidants in medicine and food. Researchers are actively investigating the bioactive chemicals found in medicinal plants for their antioxidant potential. The antioxidant effects of phytobioactives have been investigated using several bioassays.

Several researchers have made small adjustments to the radical scavenging DPPH bioassay. One of the most used methods for determining a plant extract's antioxidant activity is the DPPH test. By reacting with molecules that donate hydrogen, the stable free radical DPPH comes into play. When decreased, DPPH has a violet colour; when exposed to an antioxidant, it becomes a light yellow. Antioxidants are measured by their ability to decrease optical density. The procedure involves making a DPPH solution (0.004%) in methanol and then mixing 10 μL of the sample with 1 mL of the solution.

Both microbial infections and infectious disorders may be treated with antibiotics, which are also utilised as chemotherapeutic drugs. But very few infections really develop antibiotic resistance over time. Death, illness, and disability are all exacerbated by bacteria and viruses that are resistant to several drugs. A worldwide problem, antibiotic resistance has recently emerged. The rise of bacteria and viruses that are resistant to more than one treatment is threatening the efficacy of many existing antibiotics It emphasised the significant role of crude plant extract and phytobioactive chemicals in combating multidrug-resistant infections.

One important virulence mechanism that microbes use to survive in hostile settings is biofilm formation. On both living and nonliving surfaces, bacteria may develop biofilms. The biofilms aid microbial pathogens in infecting humans by shielding them from the effects of drugs and the immune system. Biofilms are the root cause of about 80% of bacterial illnesses in humans. Infectious disorders caused by biofilms include urethritis, vaginitis, gingivitis, conjunctivitis, colitis, and many more. *Pseudomonas aeruginosa* forms a biofilm and infects the lungs. Infections may also be caused by biofilms formed by *Escherichia coli* and *Staphylococcus aureus*. There are significant health risks associated with biofilm growth on medical equipment.

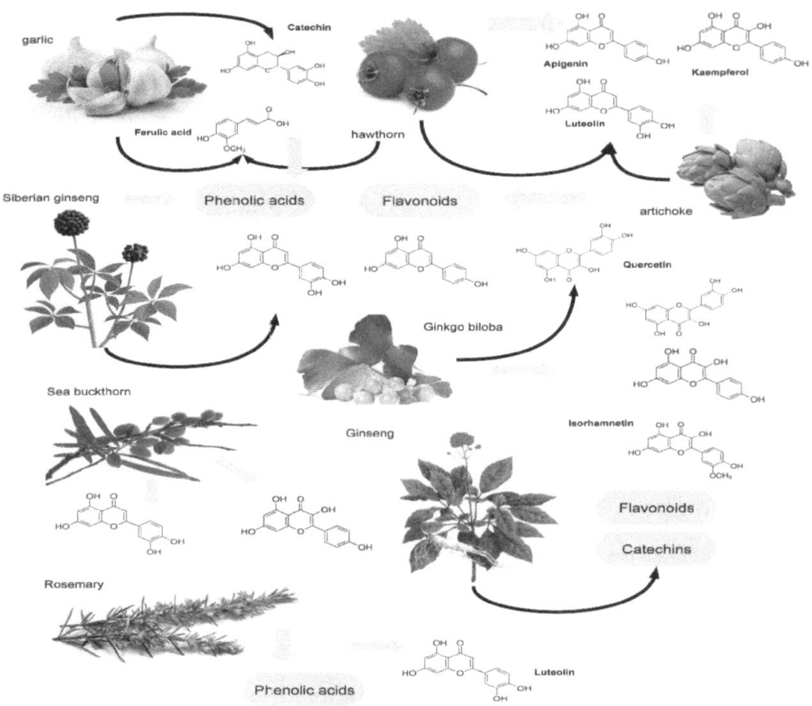

Fig 8.3 Shows the types of different bioactive compounds

The use of plant extracts to combat the spread of antibiotic-resistant bacteria has recently attracted a lot of interest. Bioactive chemicals abound in plants. Over the past few decades, researchers have studied the antibiofilm properties of a wide range of medicinal herbs.

In order to communicate with one another, bacteria of both Gram-negative and Gram-positive varieties use a mechanism called quantum sensing (QS). Bacteria use quorum sensing as a means of communication inside biofilms. Quantum sensing relies on the density of bacterial cells. It found that blocking quorum sensing might reduce bacterial pathogenicity and biofilm formation. Researchers in the medical profession are interested in exploring additional antimicrobial chemicals since they have the ability to manage microbial illnesses.

The antibacterial activities of the phytobioactive substances are very outstanding. The development of filamentous fungus may be inhibited by lipophilic chemicals derived from Moringa oleifera. These compounds attach to the cytoplasmic membrane and cause membrane permeabilization, as shown by studies. The proteins isolated from Croton tiglium L. exhibited impressive antibacterial properties. Antibacterial activity was shown to be strong in phytobioactive chemicals derived from sunflower seeds.

There are essentially two kinds of bioassays that may be used to determine whether or not natural substances have antibacterial properties. Bioassays based on cell morphology, diffusion, or bioautography may identify antimicrobial activity. The agar dilution, broth macrodilution, and broth microdilution tests are comprised of quantitative bioassays.

The isolation of phytobioactive chemicals for use as antibacterial and antioxidant agents necessitates thorough toxicity testing. Botulinum, maitotoxin, batrachotoxin, and ricin are among the world's most deadly poisons; these substances exist naturally. It is noted a significant rise in the use of in vitro bioassays for toxicity testing. Various in vitro assays have been developed for the purpose of cytotoxicity testing. These include the AMES assay, the DNA damage protection assay, the inflammatory assay, the haemolytic assay, the thrombolytic assay, and others.

The mutagenic potential of medicinal plant extract has been assessed using the AMES test. There was a standardisation effort in the 1970s to determine which chemicals may cause mutations. This assay has seen heavy application in the evaluation of herbal remedies throughout the last few years. This Salmonella typhimurium mutagenesis experiment makes use of two mutant strains: TA98 and TA100. Standard mutagens for TA98 and TA100, respectively, are potassium dichromate (K2Cr2O7) and sodium azide (NaN3). Wells that are pale yellow are thought to be positive. For a sample to be considered mutagenic, the number of positive wells in the sample plate (plant extract) must exceed the number of positive wells in the background.

Ageing and DNA damage are mostly caused by oxidative stress. Nowadays, DNA damage may be worsened by physical stress, chemical, biological, and age-related factors. To determine if plant extracts provide any protection against DNA damage, a test is used. The DNA damage prevention capability of extracts is often assessed using CT-DNA, which stands for calf thymus.

Chapter 10 Current Trends in material Physics

Muhammad Mujahid Iqbal is currently pursuing his Ph.D. at the University of Chinese Academy of Sciences (UCAS). His research focuses on the design and synthesis of novel nonlinear optical (NLO) materials, along with their structural characterization and analysis of optical properties. With a strong foundation in materials physics, his work integrates advanced techniques to explore and understand the fundamental mechanisms governing NLO behavior. He has made significant contributions to this book chapter as the main author.

9.1 Materials Physics

The field of materials physics that analyses and describes the physical characteristics of materials using physics concepts. It investigates how a material's microscopic structure influences its macroscopic characteristics, bridging the gap between fundamental physics and materials science. This area is essential for creating novel materials with specific qualities for a range of uses.

Materials scientists place a strong emphasis on comprehending how a material's processing the history affects its structure and, therefore, its characteristics and functionality. The materials paradigm is comprehension of the links between processing, structure, and attributes. This paradigm is used to enhance knowledge in a number of study fields, such as metallurgy, biomaterials, and nanotechnology.

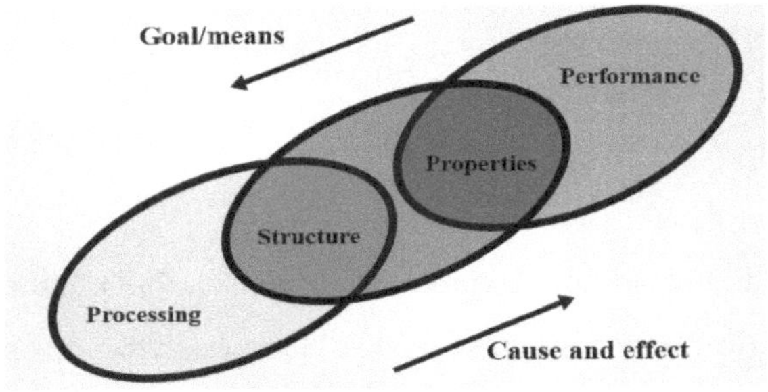

Fig 9.1 Materials physics analyses describe characteristics of materials

Material physics and metals analysis, which look into materials, goods, buildings, or components that malfunction or do not work as intended and result in property damage or human harm, also heavily rely on materials science. These kinds of studies are essential for comprehending, for instance, the reasons behind different aviation mishaps and accidents.

In terms of both quantity and economic value, iron alloys (steel, stainless steel, cast iron, tool steel, and alloy steels) account for the majority of metals in use today.

9.2 Principles of Materials Physics

Steels with low, mid, and high carbon content are made from iron alloyed with different amounts of carbon. Only when the carbon content of an iron-carbon alloy is between 0.01% and 2.00% by weight is it classified as steel. The quantity of carbon in steels determines their hardness and tensile strength; higher carbon content also results in less ductility and toughness. However, these characteristics may be considerably altered by heat treatment procedures like quenching and tempering. On the other hand, certain metal alloys have special qualities that allow their density and size to stay constant throughout a variety of temperatures. An iron-carbon alloy that has more than 2.00% but less than 6.67% carbon is called cast iron. Any ordinary steel alloy that contains more than 10% chromium by weight is referred to be stainless steel. Stainless steels usually additionally include nickel and molybdenum.

The metallic alloys of aluminium, titanium, copper, and magnesium are also important. The alloys of the other three metals were very recently produced, although copper alloys have been around since the Bronze Age. The necessary electrolytic extraction techniques were only recently developed because of the chemical sensitivity of these metals. High strength to weight ratios and, in the case of magnesium, electromagnetic shielding are further well-known and prized properties of aluminium, titanium, and magnesium alloys. These materials are perfect for the various applications in the automotive engineering sector and the aerospace industry where high strength to weight ratios are more crucial than bulk cost.

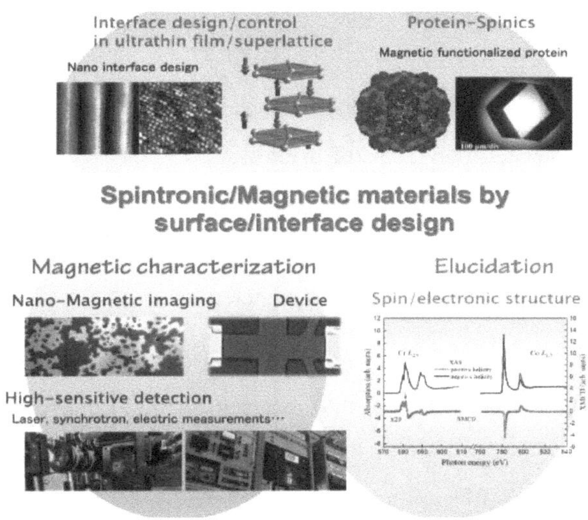

Fig 9.2 The different physical characteristics of materials

The Faraday effect and the magnetic Kerr effect are two instances of the magneto-optical effect, which is the interaction between light and a material's magnetisation. In recent years, efforts have been made to use the magneto-optical effect to optical deep neural networks, as well as to magneto-optical recording and optical isolators. A transparent magnetic material is preferred in order to take advantage of the magneto-optical effect. Nano-crystalline iron oxides like magnetite (Fe_3O_4) and haematite (Fe_2O_3) suggested as nano-crystalline materials for transparent magnetic materials.

Recently, nanogranular thin films containing magnetic nanometal particles scattered throughout a transparent matrix have been reported to exhibit a significant Faraday effect and are garnering interest. It has been reported that anisotropic properties can be exhibited by controlling the shape and dispersion of the magnetic particles, and research aimed at practical applications is also underway. Their properties are essentially isotropic and can be controlled by the materials and volume fraction of the nanomagnetic particles. On the other hand, yttrium iron garnet (YIG), a representative of rare-earth iron garnets ($RE_3Fe_5O_{12}$; REIG, RE is a rare-earth element), is a magnetic oxide material that has a large magneto-optical effect, sufficient transparency, and high magnetic permeability, strong magnetic anisotropy, low magnetic loss, and high-frequency properties.

Because of their characteristics, they might be used in the newest technical domains, such as spintronics applications, optical applications, microwave applications using ferromagnetic resonance, and spin-wave logic devices. The composition and placement of magnetic ions determine the magnetic characteristics of REIG. For example, by adding different elements to the rare-earth and iron sites, the features of REIG, such as their magnetic anisotropy, may be adjusted for certain purposes. Because of these characteristics, magnetic garnets have been the subject of much study to date.

9.3 Relationships between Structure and Properties:

The link between a material's visible qualities such as mechanical strength, electrical conductivity, and optical behaviour and its internal structure such as atomic arrangement, flaws, and interfaces is examined by materials physicists.

While statistical physics aids in describing the aggregate behaviour of many particles, quantum mechanics is utilised to comprehend the behaviour of electrons and other basic particles inside materials.

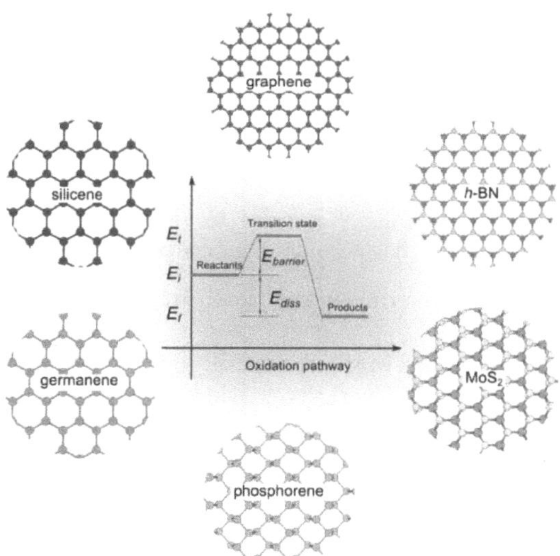

Fig 9.3 The varieties of different commonly used materials

9.4 Physics of Condensed Matter:

Superconductivity, magnetism, and other characteristics of materials in their solid, liquid, and amorphous phases are all covered in this area of physics.

Materials that are magnetic, optical, and electronic:

It investigating materials with the particular optical, electrical, or magnetic characteristics for use in photonics, electronics, and magnetics.

New Structures and Materials:

It investigating novel materials with special qualities, such 2D materials, nanomaterials, and metamaterials.

Materials' Quantum Phenomena:

It examining quantum phenomena in materials, including as topological phases, quantum entanglement, and superconductivity.

Physics in non-equilibrium:

It investigating materials that are changing quickly, as those that are being damaged by lasers or going through phase shifts.

Physics of Soft Condensed Matter:

The concentrating on substances like polymers, colloids, and liquid crystals that have intricate internal structures and characteristics.

Energy-related Materials:

The creating materials for fuel cells, batteries, and solar cells, among other sustainable energy technologies. It producing materials for use in biomedical devices such implants, biosensors, and drug delivery systems.

9.5 Applications of Materials Physics

Electronic devices:

It is creating novel materials for cutting-edge electrical devices, such as semiconductors and superconductors.

Photonics:

It producing materials for optical components such as fibres and lasers.

Magnets:

It creating materials for sensors, actuators, and magnetic storage.

Engineering Materials:

It modifying material characteristics for particular uses, such composites, coatings, and structural materials.

Nanotechnology:

It creating materials at the nanoscale for a variety of uses, including as energy storage, medicine delivery, and sensors.

Sustainable Energy:

It creating materials for energy storage devices such as batteries and solar cells.

Biomedicine:

It creating materials for tissue engineering, biosensors, and medication delivery.

References

Ahmad SS, Tahir I (2018) Putrescine and jasmonates outplay conventional growth regulators in improving postharvest performance of Iris germanica L. cut scapes. Proc Natl Acad Sci India 88:391-402

Andreou A, Brodhun F, Feussner I (2009) Biosynthesis of oxylipins in non-mammals. Prog Lipid Res 48:148-170

Arbona V, Argamasilla R, Gómez-Cadenas A (2010) Common and divergent physiological, hormonal and metabolic responses of Arabidopsis thaliana and Thellungiella halophila to water and salt stress. J Plant Phys 167:1342-1350

Bae KS, Rahimi S, Kim YJ, Devi BSR, Khorolragchaa A, Sukweenadhi J, Silva J, Myagmarjav D, Yang DC (2016) Molecular characterization of lipoxygenase genes and their expression analysis against biotic and abiotic stresses in Panax ginseng. Eur J Plant Pathol 145:331-343

Bhardwaj PK, Kaur J, Sobti RC, Ahuja PS, Kumar S (2011) Lipoxygenase in Caragana jubata responds to low temperature, abscisic acid, methyl jasmonate and salicylic acid. Gene 483:49-53

Brash AR (1999) Lipoxygenases: occurrence, functions, catalysis, and acquisition of substrate. J Biol Chem 274:23679-23682

Albert J, Fenyo EM. Simple sensitive, and specific detection of human immunodeficiency virus type 1 in clinical specimens by polymerase chain reaction with nested primers. J. Clin. Microbiol.

Baskaran N, et al. Uniform amplification of a mixture of deoxyribonucleic acids with varying GC content. Genome Res. 1996;6:633-638.

Blanchard MM, Taillon-Miller P, Nowotny P, Nowotny V. PCR buffer optimization with uniform temperature regimen to facilitate automation. PCR Methods Appl. 1993;2:234-240

Shiio, Y., Aebersold, R., Quantitative proteome analysis using isotope-coded affinity tags and mass spectrometry. Nat. Protoc. 2006, 1, 139–145.

Goodlett, D. R., Keller, A., Watts, J. D., Newitt, R. et al., Differential stable isotope labeling of peptides for quantitation and de novo sequence derivation. Rapid Commun. Mass Spectrom. 2001, 15, 1214–1221.

Welsh, JB, et al. Analysis of gene expression identifies candidate markers and pharmacological targets in prostate cancer. Cancer Res. 2001. 61:5974-5978.

Thomas, G.V., and Loda, M. 2002. Molecular staging of prostate cancer. In Prostate cancer principles & practice. P.W. Kantoff, P.R. Carroll, and A.V. D'Amico, editors. Lippincott Williams & Wilkins. Philadelphia, Pennsylvania, USA. 287–303.

DeMarzo, AM, Nelson, WG, Isaacs, WB, Epstein, JI. Pathological and molecular aspects of prostate cancer. Lancet. 2003. 361:955-964.

H. Burgsteiner, M. Kandlhofer, and G. Steinbauer, "iRobot: Teaching the Basics of Artificial Intelligence in High Schools," in 6th Symposium on Educational Advances in Artificial Intelligence, Phoenix, USA, 2016.

E. Myberg, "The effect of formal teacher education on reading achievement of 3rd-grade students in public and independent schools in Sweden," Educational Studies, vol. 33, no. 2, pp. 145–162, 2007.

A. A. Genlott and A. Gronlund, "Improving literacy skills through learning reading by writing: The iWTR method presented and tested," Computers and Education, vol. 67, pp. 98–104, 2013.

Schellenberg JC (1998) Primary Postpartum Haemorrhage. Lecture, 9th Postgraduate Course for Training in Reproductive Medicine and Reproductive Biology. Department of Obstetrics and Gynecology, Geneva University Hospital

Batista RJ, Verde J, Nery P, Bocchino L, Takeshita N, Bhayana JN (1997) Partial left ventriculectomy to treat end-stage heart disease. TA. Ann Thorac Surg 64(3):634-8

Stillwell GK (1973) The Law of Laplace: Some clinical applications. Mayo Clin Proc 48(12):863-9 PubMed CAS Google Scholar Halliday D, Resnick R, Walker J (2001) Fundamentals of physics, 6th edn. Wiley, Singapore, pp 328-329

Chintamani SV, Singh J, Bansal A, Saxena S (2005) Half versus full vacuum suction drainage after modified radical mastectomy for breast cancer- a prospective randomized clinical trial [ISRCTN24484328]. BMC Cancer 5:11 Article PubMed

William F (2005) Review of Medical Physiology 22nd International Ed. Chapter 30—Dynamics of blood & lymph flow. McGraw Hill, pp 586-587

Kumar V, Abbas Abdul K, Fausto N (2005) Robbins and Cotran pathologic basis of disease, 7th Edn, Unit 1, Chapter 5. Elsevier, pp 154-156

Z. Hrepic, P. Adams, J. Zeller, N. Talbott, G. Taggart, and L. Young, "Developing an inquiry-based physical science course for preservice elementary teachers," in 2005 Physics Education Research Conference Proceedings, 818, edited by P. Heron, L. McCollough, and J. Marx (AIP, Melville, NY, 2006), pp. 121-124.

F. Goldberg, S. Robinson, and V. Otero, Physics and Everyday Thinking (It's About Time, Herff Jones Education Division, Armonk, NY, 2007).

F. Goldberg, S. Robinson, V. Otero, R. Kruse, and N. Thompson, Physical Science and Everyday Thinking, 2nd ed. (It's About Time, Herff Jones Education Division, Armonk, NY, 2008).

L. C. McDermott, "What we teach and what is learned: Closing the gap," Am. J. Phys. 59, 301-315 (1991).

Collins, N. C., Tardieu, F. & Tuberosa, R. Quantitative trait loci and crop performance under abiotic stress: where do we stand? Plant Physiol. 147, 469-486 (2008).

Wollenweber, B., Porter, J. R. & Lubberstedt, T. Need for multidisciplinary research towards a second green revolution. Curr. Opin. Plant Biol. 8, 337–341 (2005).

Naqvi, S. et al. When more is better: multigene engineering in plants. Trends Plant Sci. 15, 48–56 (2010).

Venter, M. Synthetic promoters: genetic control through cis engineering. Trends Plant Sci. 12, 118–124 (2007).

Azzam R et al. (1983) In: Proc Intl Symp on Isot and Rad Tech in Soil Phys and Irrig Stud Aix-en-Provance, p 321

Andrade JD (ed) (1976) Hydrogels for medical and related applications. ACS symp ser Washington

Peppas NA (ed) (1986) Hydrogels in medicine and pharmacy. CRC Press, Boca Raton

Flory PJ (1953) Principles of polymer chemistry. Cornell Univ Press, Ithaca, New York

Zhang, Y.; de Boer, B.; Blom, P.W.M. Controllable molecular doping and charge transport in solution-processed polymer semiconducting layers. Adv. Funct. Mater. 2009, 19, 1901–1905.

Zhang, Y.; Blom, P.W.M. Electron and hole transport in poly(fluorene-benzothiadiazole). Appl. Phys. Lett. 2011, 98, 143504.

Nollau, A.; Pfeiffer, M.; Fritz, T.; Leo, K. Controlled n-type doping of a molecular organic semiconductor: Naphthalenetetracarboxylic dianhydride (NTCDA) doped with bis(ethylenedithio)-tetrathiafulvalene (BEDT-TTF). J. Appl. Phys. 2000, 87, 4340–4343

Bajpai, M.; Srivastava, R.; Dhar, R.; Tiwari, R.S. Review on optical and electrical properties of conducting polymers. Indian J. Eng. Mater. Sci. 2016, 2016, 1–8.

Patil, A.O.; Heeger, A.J.; Wudl, F. Optical properties of conducting polymers. Chem. Rev. 1988, 88, 183–200.

Huang, W.S.; Humphrey, B.D.; MacDiarmid, A.G. Polyaniline, a novel conducting polymer. Morphology and chemistry of its oxidation and reduction in aqueous electrolytes. J. Chem. Soc. Faraday Trans. 1 1986, 82, 2385–2400.

McCullough, R.D.; Lowe, R.D.; Jayaraman, M.; Anderson, D.L. Design, synthesis, and control of conducting polymer architectures: Structurally homogeneous poly(3-alkylthiophenes). J. Org. Chem. 1993, 58, 904–912.

Wang, J.; Dai, J.; Yarlagadda, T. Carbon nanotube conducting polymer composite nanowires. Langmuir 2005, 21, 9–12.

Lee, B.H.; Lee, J.H.; Kahng, Y.H.; Kim, N.; Kim, Y.J.; Lee, J.; Lee, T.; Lee, K. Graphene-conducting polymer hybrid transparent electrodes for efficient organic optoelectronic devices. Adv. Funct. Mater. 2014, 24, 1847–1856

Gupta, S.; Price, C.; Heintzman, E. Conducting polymer nanostructures and nanocomposites with carbon nanotubes: Hierarchical assembly by molecular electrochemistry, growth aspects and property characterization. J. Nanosci. Nanotechnol. 2016, 16, 374–391.

Cantwell, W.J.; Morton, J. The impact resistance of composite materials—A review. Composites 1991, 22, 347–362.

A. Manthiram, in Lithium Batteries: Science and Technology (Eds: G. A. Nazri, G. Pistoia), Springer, New York 2009.

De Sagrera JE (2004) Dali científico Las ciencias biomédicas en la obra del pintor ampurdanés. Offarm 23:122–128

Gonçalves-Maia R (2017) Linus Pauling. A natureza da ligação química. Edições Colibri, Lisbon. ISBN 978-989-689-692-8.

Mirsky AE, Pauling L (1936) On the structure of native, denatured, and coagulated proteins. Proc Natl Acad Sci U S A 22:439–447.

Corey RB, Pauling L (1950) Two-hydrogen spiral configurations of the polypeptide chain. J Am Chem Soc 72:5349–5349.

Pauling L et al (1951) The structure of proteins: two hydrogen-bonded helical configurations of the polypeptide chain. Proc Natl Acad Sci U S A 37:205–211.

Pauling L, Corey RB (1951) Configurations of polypeptide chains with favored orientations around single bonds: two new plated sheets. Proc Natl Acad Sci U S A 37:729–740.

Pauling L, Corey RB (1953) A proposed structure of the nucleic acids. Proc Natl Acad Sci U S A 39:84–97.

Angchuk P, Keller PA, Pyne SG, Taweechotipatr M, Tonsomboon A, Rattanajak R, et al. Screening of ethnopharmacologically selected Bhutanese medicinal plants for their phytochemical profiles and biological activities. J Ethnopharmacol. 2011;137(1):730–42.

Cullen DR, Pengon J, Rattanajak R, Chaplin J, Kamchonwongpaisan S, Mocerino M. Scoping studies into the structure-activity relationship (SAR) of phenylephrine-derived analogues as inhibitors of Trypanosoma brucei rhodesiense. ChemistrySelect. 2016;1:4533–8.

Collins L, Franzblau SG. Microplate alamar blue assay versus BACTEC 460 system for high-throughput screening of compounds against Mycobacterium tuberculosis and Mycobacterium avium. Antimicrob Agents Chemother. 1997;41(5):1004–9.

Nguyen MTT, Nguyen HX, Huynh TNN, Min BS. Screening of α-glucosidase inhibitory activity of Vietnamese medicinal plants: isolation of active principles from Oroxylum indicumin. Nat Prod Sci. 2012;18:47–51.

Xiao Z, Storms R, Tsang A. Quantitative starch-iodine method for measuring alpha-amylase and glucoamylase activities. Anal Biochem. 2006;351(1):146–8.

Lin LJ, Huang XB, Lv ZC. Isolation and identification of flavonoids components from Pteris vittata L. SpringerPlus. 2016;1(1):1649.

Kashiwada Y, Iizuka H, Yoshioka K, Chen R, Nonaka G, Nishioka I. Tannins and related compounds. XCIII. Occurrence of enantiomeric proanthocyanidins in the Leguminosae plants, Cassia fistulan L. and Cassia javanica L. Chem Pharm Bull. 1990;4:888-93.

Shi T, Chen H, Jing L, Liu X, Sun X, Jiang R. Development of a kilogram-scale synthesis of salidroside and its analogs. Synth Commun. 2011;177(17):2594-600.

Truong D, Nguyen DH, Ta NTA, Bui AV, Do TH, Nguyen HC. Evaluation of the use of different solvents for phytochemical constituents, antioxidants, and in vitro anti-inflammatory activities of Severinia buxifolia. J Food Qual. 2019;8178294:1-9.

Rai KN, Kaushalendra K, Singh J. Chemical constituents from the pods of Cassia glauca Lam. Asian J Chem. 1997;1:44-9.

Tiwari HP, Misra M. Phytochemical investigation of Cassia glauca bark. J Ind Chem Soc. 1993;7:653.

Rai KN, Roy RA. A new anthraquinone glycoside from the stem of Cassia glauca, Lam. J Bangladesh Acad Sci. 1991;2:193-8. Google Scholar Hemlata S, Kalidhar SB. Chemical components of Cassia glauca Lam. Ind J Pharma Sci. 1994;1:33-4.